GIS for Environmental Stewardship and Streamlining

An Overview of State DOT Practices

Prepared by the Volpe National Transportation System Center for the Federal Highway Administration Office of Project Development and Environmental Review

January 2005

GIS for Environmental Stewardship and Streamlining
An Overview of State DOT Practices

GIS for Environmental Stewardship and Streamlining 1
Introduction 3
Applying GIS to Transportation Decisionmaking 4
Methodology 5
Overview of Cases 6
Data Management 6
    Indiana 6
Cooperation and Outreach 8
    Arkansas 8
    Ohio 10
    Washington 11
Spatial Analysis and Modeling 13
    Minnesota 13
    Texas 15
Re-engineering DOT Business Practices 16
    Virginia 16
    Florida 17
    GIS4EST Catalysts 19
Lessons Learned and Recommendations 20
    Look Beyond Initial Costs 20
    Start small and build as you go 21
    Collaborate and Build Partnerships 22
    GIS4EST Applications to Meet Local Streamlining Challenges 23
    Focus on Evaluation Measures From the Outset 23
Appendix A: Interview Guide 24
Appendix B: Indiana Case Study 25
Appendix C: Arkansas Case Study 29
Appendix D: Ohio Case Study 32
Appendix E: Washington Case Study 36
Appendix F: Minnesota Case Study 40
Appendix G: Texas Case Study 44
Appendix H: Virginia Case Study 48
Appendix I: Florida Case Study 52

GIS for Environmental Stewardship and Streamlining

An Overview of State DOT Practices

INTRODUCTION

Emerging technologies are transforming the way State Departments of Transportation (State DOTs) plan and implement new transportation infrastructure and improvements to existing networks. One of the most promising innovations in recent decades is Geographic Information Systems (GIS), a technology that—among other uses—has great potential to improve transportation decisionmaking. State DOTs are increasingly adopting GIS technologies to promote better environmental stewardship, while concurrently streamlining the environmental review process required by the National Environmental Policy Act (NEPA) and other environmental laws and regulations.

Passed in 1969, NEPA requires Federal agencies to conduct environmental reviews on any action that might "significantly affect the quality of the human and natural environment." NEPA has fundamentally changed the Federal structure for planning and project development, and has resulted in important environmental protections. However, over the past several decades an increasing number of policymakers, civil servants, and citizens have voiced concern over the amount of time required by the environmental review process, which has increased from an average of 2.2 years in the 1970s to 5 years in the 1990s[1] and to 6 years in FY2004.[2] As a result, Section 1309 of the 1998 Transportation Equity Act for the 21st Century (TEA-21) addressed the need to shorten the overall timeframe for the project development process while continuing to safeguard environmental and cultural resources. In addition, in 2002 President Bush signed Executive Order (EO) 13274, *Environmental Stewardship and Transportation Infrastructure Project Reviews*, "to enhance environmental stewardship and streamline the environmental review and development of transportation infrastructure projects."

The Federal Highway Administration (FHWA) supports the adoption and development of GIS technologies to promote environmental streamlining and stewardship (from here on referred to as "GIS4EST"). GIS4EST also supports FHWA's Environmental Vital Few Goal,[3] which is fundamentally aimed at improving project delivery without compromising environmental protection. One specific objective of the Environmental Vital Few Goal is to integrate transportation decisionmaking with multimodal planning, the environmental process, and project development at a systems level. Another objective is to decrease the median time it takes to complete an Environmental Impact Statement (EIS) from 54 months to 36 months and the median time to complete an Environmental Assessment (EA) from approximately 18 months to 12 months by September 30, 2007. FHWA has identified GIS technologies, when properly applied and managed, as a tool to achieve these objectives.

GIS4EST is also a natural outgrowth of several of the recommendations made to the Council on Environmental Quality (CEQ) in the NEPA Task Force's 2003 report, *Modernizing NEPA*. Two recommendations were:

---

[1] See *Evaluating the Performance of Environmental Streamlining: Development of a NEPA Baseline for Measuring Continuous Performance*. Written by the Berger Group for FHWA. Available at http://www.environment.fhwa.dot.gov/strmlng/baseline/index.htm

[2] See *Estimated Time Required to Complete the NEPA Process*, compiled by FHWA Office of Project Development and Environmental Review. Available at http://www.environment.fhwa.dot.gov/strmlng/nepatime.htm

[3] For more information, see http://environment.fhwa.dot.gov/strmlng/vfovervw.htm.

GIS for Environmental Stewardship and Streamlining
An Overview of State DOT Practices

Clarify appropriate roles of communications and information dissemination technologies in the NEPA process to enhance public involvement.

Coordinate with interagency groups about protocols and standards pertaining to data, information management, modeling tools, and information security.

While several States have already applied or are in the process of adopting GIS4EST, other States are just beginning down this path. Building on information collected through interviews with State DOT officials and GIS specialists, this report highlights the way eight State DOTs are using GIS to promote environmental stewardship and streamlining, revealing both the potential of GIS4EST and its varied applications.

APPLYING GIS TO TRANSPORTATION DECISIONMAKING

A GIS is a collection of computer software, hardware, data, and personnel used to store, manipulate, analyze, and present geographically referenced information. Users store spatial features in a spatial database, or a coordinate system that references the Earth. Users can then associate attribute (also known as "tabular" or "descriptive") data with these spatial features and layer spatial data with its associated attribute information for viewing and analysis. Using GIS, multiple items of interest about a particular geographic area can be displayed and analyzed.

Both governmental and non-governmental institutions are adopting GIS technology as spatial data become more widely available. Many sources of data are now available on the World Wide Web for little or no cost. In addition, while GIS still requires some special training, the technology is becoming considerably more user-friendly. For example, Internet Mapping Servers (IMS) offer a way to provide mapping capabilities to the public in a way that involves little or no training and does not require each user to own expensive GIS software. As a result of these innovations to make GIS more user-friendly and accessible, increasing numbers of institutions are developing their own spatial data and GIS applications tailored to meet local needs.

As applied to environmental stewardship and streamlining goals, GIS technologies advance several of the Enlibra principles conceived by Utah Governor Mike Leavitt and Oregon Governor John Kitzhaber.[4] Together, the eight Enlibra principles form a philosophy based on a balanced approach to environmental management and a shared commitment to stewardship of the environment. The use of GIS4EST has been a key component of initiatives to integrate three of these principles into business processes:

<u>Collaboration, not Polarization</u> – *Use collaborative processes to break down barriers and find solutions.* GIS4EST promotes collaboration by enabling stakeholders to come together to identify data needs and gather data. GIS4EST also provides a neutral forum for discussion of emerging issues and enables State DOTs to solicit more meaningful input from cooperating agencies and the public.

<u>Recognition of Benefits and Costs</u> – *Make sure all decisions affecting infrastructure, development, and the environment are fully formed.* By allowing decisionmakers to reference

---

[4] *Enlibra* literally translates as "to move towards balance." For more information about the *Enlibra* principles, see http://www.oquirrhinstitute.org/em_leavitt.html

GIS for Environmental Stewardship and Streamlining
An Overview of State DOT Practices

a database of information that is maintained and updated over time, GIS4EST promotes better decisionmaking about potential impacts to environmental and cultural resources and allows agencies and the public to more explicitly portray the tradeoffs involved in different project alternatives.

Science for Facts, Process for Priorities- *Separate subjective choices from objective data gathering.* An integrated spatial database can serve as the foundation for streamlined decisionmaking ensuring that all stakeholders base their analysis, judgments, and opinions on the same information. GIS4EST practices can foster an enhanced understanding of project decisions, especially when all stakeholders are involved in validating data and articulating assumptions. In particular, GIS technologies can provide a more complete assessment of cumulative impacts—an issue that is emerging as a common stumbling block in the environmental review process.

METHODOLOGY

The information contained in this report was collected through interviews with State GIS specialists and project managers who manage GIS4EST work. With consultation from FHWA headquarters, eight geographically diverse States were selected for interviews.

The GIS applications developed by these States represent the full spectrum of GIS development. The GIS applications that State DOT officials described to us fall into four general categories of use:

- Data management
- Interagency coordination
- Spatial analysis and modeling
- Re-engineering business processes

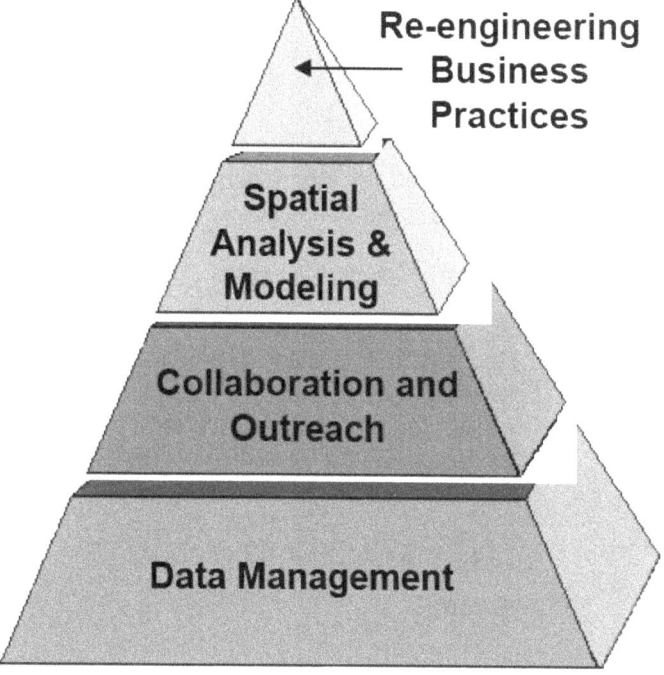

While State DOTs vary in the process by which they adopt and apply GIS4EST technologies, these categories suggest a rough sequencing for the development of a GIS4EST applications from less to more complex systems. These applications are discussed below, grouped by the phase of development that best describes their GIS4EST efforts to date.

Data Management

- INDIANA: GIS for Southwest Indiana, GIS Atlas for Indiana

GIS for Environmental Stewardship and Streamlining
An Overview of State DOT Practices

Interagency Cooperation

- ARKANSAS: GIS for the Interstate-69 Southeast Arkansas Connector
- OHIO: GLO Map Digitization Project, Cultural Resources GIS
- WASHINGTON: The Environmental GIS Workbench

Spatial Analysis and Modeling

- MINNESOTA: Mn/Model
- TEXAS: GIS Screening Tool (GISST)

Re-engineering Business Practices

- VIRGINIA: Enterprise GIS, Natural Heritage Resource Database, and Comprehensive Environmental Data and Reporting (CEDAR)
- FLORIDA: The Environmental Screening Tool (EST)

Volpe staff conducted these interviews between January and May 2004. Additional material about certain applications was compiled via Internet research and from subsequent correspondence and conversations with State DOT staff.

OVERVIEW OF CASES

This section provides highlights of these eight cases, providing an overview of the development of GIS4EST applications in each State. The cases below present information on the challenges and solution States faced in the development process, the environmental stewardship and streamlining accomplishments of the GIS4EST applications, as well as next steps in furthering their GIS4EST applications.

DATA MANAGEMENT

Indiana

The Indiana Department of Transportation (INDOT) has compiled over 170 layers of information commonly used for planning and environmental purposes to create a statewide GIS. Entitled *A GIS Atlas for Indiana*, this statewide system was designed to foster the consideration of potential environmental impacts early in the project development process.

Challenges and Solutions

The project had its inception in 1999 when INDOT began the environmental review process for the Interstate 69 (I-69) project in southwestern Indiana, a roadway that will span 142 miles in Indiana and a study area of 26 counties.[5] I-69 is also known as the "NAFTA

---

[5] The Final EIS was released on the project in December of 2003 and is available at http://www.in.gov/dot/projects/I69/.

GIS for Environmental Stewardship and Streamlining
An Overview of State DOT Practices

highway"—a congressionally mandated, 1,600-mile interstate highway stretching from the Mexican border in Brownsville and McAllen, Texas to the US-Canadian border in Detroit, Michigan. I-69 was chosen in 2002 as a streamlining pilot project[6] under TEA-21, Section 1309.

INDOT, with the assistance of an engineering and environmental consulting firm, utilized a tiered environmental document for the project. Recognizing since they were going to collect environmental data over such a large area, INDOT sought a way to permanently store this work for future use. INDOT identified the development of a GIS as the best way to accomplish this goal. INDOT and its consultants, therefore, subcontracted with the Indiana Geological Survey (IGS) of Indiana University to compile the GIS for Southwest Indiana,[7] a system that now contains 173 layers of geospatial information.

The GIS for Southwest Indiana was intended to be a project-specific GIS for the I-69 corridor. However upon its completion, INDOT could clearly see the value of expanding the spatial database to the entire State. In April 2002, INDOT granted IGS funds to expand the GIS into a statewide computer-based atlas. Entitled the GIS Atlas for Indiana,[8] this statewide project now contains 206 layers of free downloadable spatial data.

The main features of the GIS Atlas for Indiana website include an Interactive Mapping Server (IMS) enabling the construction of maps via the Internet; downloadable files of spatial data, including reference, demographic, infrastructure, environmental hydrologic, and geologic data; and metadata text files for each data layer.

Environmental Stewardship and Streamlining

The cost of the GIS Atlas for Southwest Indiana was about $100,000, a fraction of the cost of reaching a typical Record of Decision (ROD) which INDOT officials estimate usually cost around $500,000. The GIS Atlas for Indiana has been funded through a 3-year State Planning and Research (SPR) grant of $850,000, which INDOT is matching at 20 percent. INDOT anticipates that after two or three environmental reviews, the time- and money-savings generated from the GIS databases will have paid off the up-front costs of developing these systems.

---

[6] For a complete list of the priority projects, go to http://www.fhwa.dot.gov/stewardshipeo/pplist.htm.

[7] Available at http://igs.indiana.edu/arcims/southwest/viewer.htm

[8] Available at http://igs.indiana.edu/arcims/statewide/index.html

GIS for Environmental Stewardship and Streamlining
An Overview of State DOT Practices

The GIS Atlas for Indiana saves time and money by minimizing the need for information to be "chased down." Sensitive resources can be avoided early on, when the greatest flexibility in terms of avoiding impacts exists. INDOT's consultants note, "the number one principle of mitigation is to avoid bad projects," and GIS has been particularly useful because it helps them visualize potential impacts early. INDOT's consultants note that whenever a firm works on an EIS using the GIS, they have better quality data, can get to "common ground" more quickly, and save time and money. Having a commonly referenced set of data that is up-to-date has both reduced inter-agency and public conflict, and promoted better environmental decisions.

The contractors working on the I-69 project also cite the GIS4EST as particularly useful in building credibility with the public on EISs. Providing the public with the same data that are being used to evaluate alternative alignments minimizes unnecessary conflict over "what's out there" and helps build consensus on a Locally Preferred Alternative more quickly.

Next Steps

While INDOT notes their preference to partner with other organizations to develop the GIS Atlas for Indiana into a statewide consortium, INDOT is also willing to shoulder future maintenance costs themselves, if necessary, because the cost savings to the transportation agency are so substantial.

COOPERATION AND OUTREACH

Arkansas

Since the early 1990s, the Arkansas Highway and Transportation Department (AHTD) has been developing GIS applications to support the analysis of alternative project alignments, with a growing degree of internal staff and consultant expertise and comfort with the technology. Through this incremental growth, AHTD recognized that GIS could help determine project impacts for EISs in a more efficient manner. In AHTD's view, the technology could provide a quick, accurate, and precise instrument for the generation of maps detailing the environmental constraints for multiple alternative alignments for a proposed project.

One of AHTD's earliest GIS efforts was digitizing General Land Office (GLO) survey maps from the 18th and 19th centuries. These maps reveal features that have since disappeared from the visible landscape—such as historic streambeds, Native American mound sites, and homesteads. Knowing where these unique features are helps transportation planners avoid or minimize impact to these features.

Challenges and Solutions

Like INDOT, AHTD developed a GIS4EST related to the I-69 NAFTA highway. AHTD utilized GIS technology to streamline the transportation decisionmaking and permitting process for the Southeast Arkansas Connector (I-69 SE-Connector). The I-69 Connector is the first of three I-69 projects in Arkansas and will ultimately connect Interstate 69 to the existing Arkansas interstate highway system. GIS enabled AHTD to share and consolidate environmental and engineering data. The technology also allowed large amounts of study area information to be refined and efficiently analyzed.

GIS for Environmental Stewardship and Streamlining
An Overview of State DOT Practices

Another key use of AHTD's GIS on the I-69 SE-Connector was to foster early coordination with resource agencies, the public, and Native American tribes while efficiently addressing the requirements of the environmental review process. AHTD delineated two-mile wide preferred corridors, each with 300-foot alignments. GIS coverages containing environmental constraint data were overlaid on each of the preferred corridors, allowing for quick and thorough identification of Draft EIS alternatives. 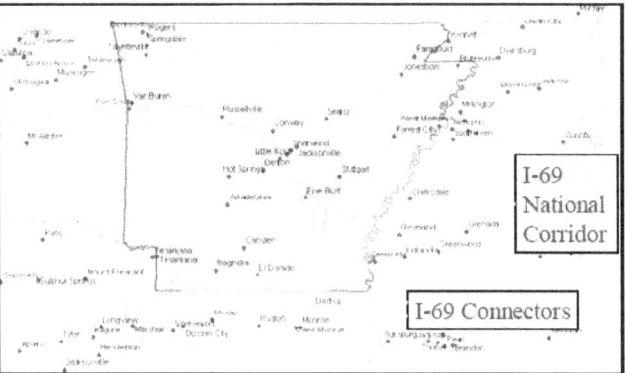 The GIS-generated maps and analyses provided partnering agencies and communities tangible examples of how various project alternatives would impact environmental, cultural, and economic resources. Partnering agencies supported GIS use because project steps occurred more quickly. The public especially welcomed the GIS and appreciated the map visualizations. AHTD noted that public participants were eager to learn how the project would affect their neighborhoods, properties, and houses. By providing this information, AHTD was able to garner quick public response on the subtle differences of proposed alternatives.

Environmental Stewardship and Streamlining

Costs for developing, implementing, and maintaining AHTD's GIS were not trivial; however, the payoff in timesavings is commensurate to their investment. Initial costs of GIS program implementation were roughly $100,000. In addition, AHTD maintains a small GIS unit with a staff of five. AHTD measures the benefits of the GIS in terms of timesavings on an environmental impact assessment. Previously, the time required to move from the Notice of Intent (NOI) to the ROD averaged 62 months. For the 1-69 SE Connector project, the ROD was signed in 26 months, a 58 percent reduction in time. Though the expedited schedule was due in part to efforts to coordinate with other agencies early and often, AHTD attributes a majority of the timesavings to the use of this GIS4EST application.

Next Steps

Currently, AHTD is evaluating other ways to use GIS to streamline the NEPA process for projects requiring an EIS. The DOT is conducting an archaeological survey to compile all of Arkansas' archaeological data in a web-based GIS format. The GIS will be accessible to archaeologists and transportation professionals on an intranet system. AHTD expects the system to aid in tribal consultation. Similarly, AHTD is also currently developing a historic bridge management GIS.

AHTD and the FHWA Arkansas Division hope to expand their use of GIS4EST by participating in the Federal GIS Users Group, a consortium of Federal agencies within Arkansas that share geospatial data. Group meetings present a forum for agencies to share GIS data, discuss projects that may affect other participating agencies, and reduce duplications in data gathering.

GIS for Environmental Stewardship and Streamlining
An Overview of State DOT Practices

Ohio

In the late 1990s, the Ohio Department of Transportation (ODOT) Office of Environmental Services (OES) formed an innovative alliance with the Ohio Historical Society/Ohio State Historic Preservation Office (OHS/OSHPO) to develop a GIS based on Mapping and Preservation Information Technology (MAPIT)[9] software to document over 120,000 Ohio Historic Inventory (OHI) and Ohio Archaeological Inventory (OAI) features. These features include individual properties and historic districts listed on the National Register of Historic Places in Ohio. Because OSHPO needed additional funding to develop such a system, ODOT agreed to help finance this project. The joint development of the GIS was a win-win situation for both agencies; OSHPO obtained the resources they needed to thoroughly systematize cultural resources in the State, and ODOT gained electronic access to the newly formed cultural resource database at OSHPO. These data inform the way ODOT develops potential transportation alignments and are invaluable for planning purposes, as well as for the NEPA process in general.

Challenges and Solutions

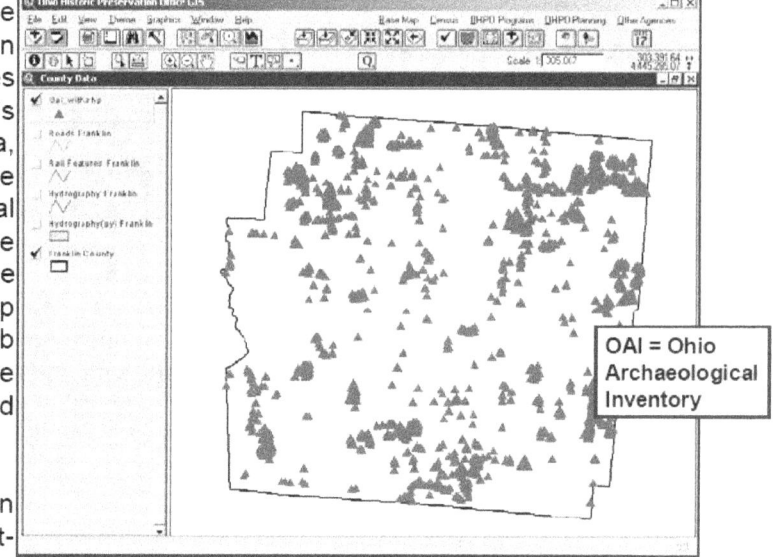

The primary challenge in developing the GIS was to create an interface between the different GIS software packages utilized by OSHPO and ODOT (ESRI's ArcView and Intergraph's GeoMedia, respectively). Both agencies were reluctant to invest the time and financial resources necessary to convert to the other agency's GIS software. In fact, the original concept about how to develop this system (putting data on MAPIT web browser) was not workable because ODOT was disinclined to purchase and deploy a new system.

ODOT and OSHPO developed a solution to this problem: create a third Internet-based system to act as an intermediary between MAPIT and GeoMedia. The use of this intermediary system obviates the need for either agency to purchase an entirely new GIS software package or for staff to invest time in learning a new system. With the current configuration, ODOT staff can use the GeoMedia software on their desktop computers to read OSHPO's ArcView data via the Internet portal.

Environmental Stewardship and Streamlining

ODOT's and OSHPO's GIS4EST application contributes to environmental stewardship and streamlining goals in several ways. One way is by strengthening working relationships between the DOT and the SHPO. In fact, the very process of creating the GIS has

---

[9] *The Mapping and Preservation Inventory Tool* (MAPIT) is a National Park Service adaptation of ESRI's ArcView. MAPIT organizes historic resource inventories in a computerized database with sophisticated mapping capabilities. See http://www.conservationgis.org/links/historic.html or http://www.cr.nps.gov/map.htm.

GIS for Environmental Stewardship and Streamlining
An Overview of State DOT Practices

strengthened interagency working relationships and contributed to a high level of trust between ODOT and OSHPO. By meeting regularly with the OSHPO, ODOT has benefited with more timely review and approvals and a better understanding of the agency's goals and objectives in regards to the environment.

Transportation decisionmakers frequently cite Section 106 as one of the most cumbersome and contentious issues in the environmental review process. Yet according to ODOT officials "it's rare for Section 106 to be a hot button issue now—in other words to determine whether or not a project will be delivered on time." ODOT credits good working relationships and the GIS with its relatively smooth track record in complying with Section 106 requirements.

The screening capabilities of the GIS4EST have also lead to time- and cost-savings. With the GIS, detailed information can be queried or "boiled down" and displayed visually for use in the early stages of transportation project planning. This process enables OES staff to electronically plot archaeological and historic site locations and make early evaluations of potential impacts in a transportation project area. GIS affords ODOT staff the opportunity to screen an area for obviously significant cultural resources and therefore devote resources and detailed analysis only to the sites that truly warrant it.

Lastly, the GIS4EST MAPIT program eliminates the time ODOT staff and consultants used to spend driving to the Ohio Historical Preservation Office to manually look through files to extract and copy information. ODOT has already saved hundreds of hours in data collection, thus leaving time for staff to spend on other activities and enable project reviews to move more quickly.

Next Steps

ODOT has completed a previously surveyed areas project with OSHPO for digitizing areas subjected to archaeological survey for any kind of initiative (e.g., housing, pipeline, roadway). The purpose of this project is to quickly determine the extent of a survey area, as well as to obtain references to reports and inventory forms. In the future, ODOT officials have expressed an interest in undertaking a similar project for historic and architectural resources. ODOT also has plans to integrate historic bridge locations embedded in their GIS system.

In addition, ODOT is currently funding two initiatives with OSHPO. The first is to build a database of properties determined eligible for the NRHP through the Section 106 review process. ODOT officials believe that this will be an invaluable database because that information is the hardest to locate in OSHPO's files as it is primarily in consultation letters between agencies. ODOT is also undergoing a project with them to scan all the NRHP, OAI, and OHI forms (approximately 140,000 total) so ODOT will have electronic versions of all those forms linked to the GIS database. When this is complete, ODOT will be able to click on a property location and pull up its complete inventory form.

Washington

The Environmental GIS Workbench (Workbench) is a custom application built to assist Washington Department of Transportation (WSDOT) staff in accessing over 60 layers of environmental, natural resource management, and transportation data. The WSDOT Environmental Information Program works with Federal, State, and local agencies to maintain a collection of the best available data for statewide environmental analysis.

GIS for Environmental Stewardship and Streamlining
An Overview of State DOT Practices

Challenges and Solutions

The Workbench provides WSDOT staff with a tool to locate proposed transportation projects and display relevant environmental data themes for that location or route. Prior to the Workbench, users seeking this data had to navigate through a difficult environment that required them to know detailed information about scale and agency management of the data. The Workbench now provides a more intuitive method to locate GIS data on four servers. Users now have the capability to build custom maps in real time, perform spatial analysis, and create hardcopy prints of their work. In addition, central management of data by expert staff improves data quality and reduces data redundancy throughout the State.

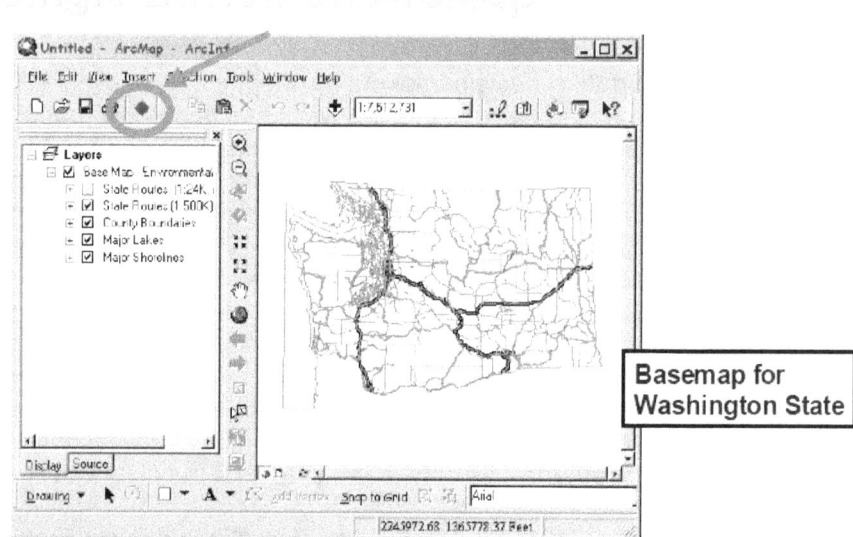

The first Workbench is an ArcView 3x extension that was written in Avenue code. To users, the Workbench appears as an extra blue button on the top tool bar in ArcView.

Environmental Stewardship and Streamlining

WSDOT characterizes the Workbench as an excellent return on their investment. The Environmental Information Program finds that training staff to use the Workbench to make basic maps and retrieve spatial information saves GIS staff time in the long-run as other staff no longer rely on them for these tasks. As a result, the Environmental Information Program staff can focus their energy and time on managing and collecting new data. Another reason why the Workbench has been cost effective is because it is user-friendly and, therefore, has not required staff to undergo extensive training to acquire many additional skills.

The impact that the Workbench has had on the scoping process is significant. While the Environmental Information Program staff has not conducted a formal evaluation, they do note anecdotal evidence of time and cost savings. The GIS has been of primary importance in saving research time. For instance, one project manager reported that the scoping process, which used to take eight hours, could be completed in two hours with the Workbench. This anecdotal evidence suggesting that the Workbench reduces scoping time by 75 percent indicates that the application has already paid for itself.

The Workbench has also been useful on specific environmental issues. WSDOT credits having a standard dataset available and ready to use as a particular asset in negotiating disputes involving endangered species, a strongly contested issue in Washington State. In particular, WSDOT is using the Workbench to set acceptable limits and times on construction scheduling due to concerns about endangered species.

GIS for Environmental Stewardship and Streamlining
An Overview of State DOT Practices

Next Steps

A new tool is being developed for the Workbench that will incorporate land use land cover, geology, soils, wetlands, floodplains, steep slopes, critical aquifers, parks, hydrography, and existing transportation infrastructure as inputs to a spatial model that will generate a mitigation risk index. The mitigation risk index estimates the cost effectiveness of mitigation highway impacts within the ROW. The more negative the resulting value, the stronger the implication that existing conditions and characteristics of the proposed project area will have a difficult time creating on-site, in-kind mitigation. Positive values indicate that mitigation is feasible within the ROW. A "perfect" score of 1.0 indicates conditions favorable to minimizing mitigation costs. The GIS tool will have an interface for users to input project locations, answer a few questions about project activities, and review the list of data that will be used. Once the inputs are validated, the model runs and provides the user with some statistics and a related explanation regarding mitigation issues.

While the initial purpose of the Workbench was to support project scoping, planning, and engineering, permitting staff have also become interested in using the application. Because of this growing interest, the scope of the Workbench itself is expanding. The next generation Workbench is expanding the utility of the application beyond solely environmental purposes to a new focus on maintenance and transportation planning. Elizabeth Lanzer, the Environmental Information Program Manager, describes this as an effort to expand the usefulness of the application "to meet all the GIS business needs of WSDOT."

SPATIAL ANALYSIS AND MODELING

Minnesota

Since 1996, the Minnesota Department of Transportation (Mn/DOT) has been developing an archaeological predictive model, Mn/Model, to avoid impacts to archaeological sites throughout Minnesota. An archaeological predictive model is a tool that indicates the probability of encountering an archaeological site anywhere within a given area. Using these models, construction projects can be modified to avoid areas where archaeological sites are likely to be present. The goal of Mn/Model is to be accurate enough to predict 85 percent of known archaeological sites without designating more than 33 percent of the State's area as high or medium site probability. MnDOT used GIS and statistical analysis to produce the current archaeological predictive model so that it could be replicated by anyone using the same data and following the same procedures.

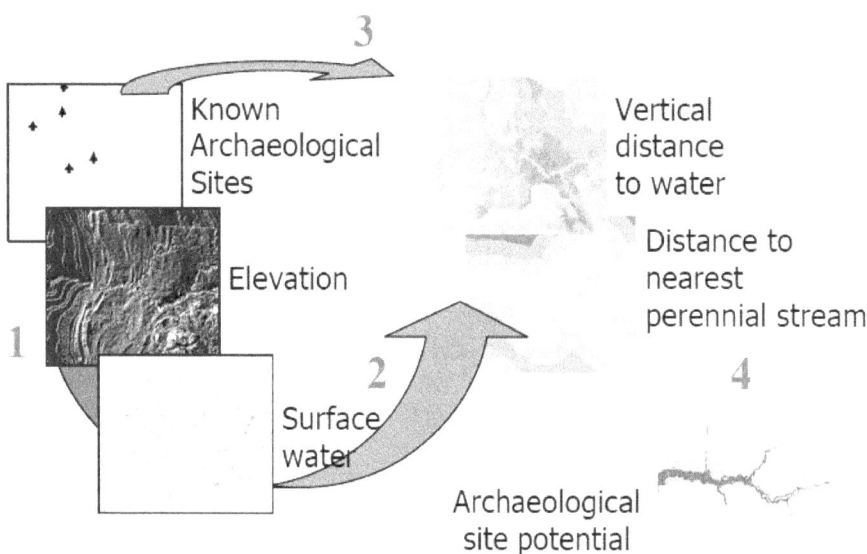
© Minnesota Department of Transportation, 2001

13

GIS for Environmental Stewardship and Streamlining
An Overview of State DOT Practices

Mn/DOT uses the predictive model for site avoidance and survey design. The results of Mn/Model are incorporated into the earliest phases of project planning, alerting transportation planners to the possible locations of archaeological sites. Mn/Model allows planners to prepare alternative avoidance design scenarios, when possible, and to budget for survey and mitigation costs and time when avoidance is not possible. Mn/Model also helps prepare budget and schedule estimates allotted for individual projects and longer range management activities.

Challenges and Solutions

Previously, Mn/DOT hired expert SHPO archaeologists to develop surveys required by Section 106 of the National Historic Preservation Act of 1966. The survey methods employed depended on the experience of the professionals involved and changed frequently. Mn/DOT, frustrated with the time and costs associated with survey and subsequent review, decided to develop a tool that would enable staff to predict where archaeological resources are likely to be found.

The probability of finding cultural resources sites are reflected in sensitivity maps. These maps usually contain three zones: a high sensitivity area where archaeological sites are most likely, a medium sensitivity area where sites are less likely, and a low sensitivity area where sites are unlikely. The model was run in 1998 and calculated a site's likelihood of containing an archaeological site by analyzing 70 spatial environmental variables, such as terrain, proximity to water, slope, and vegetation. The output from the model was displayed as maps dividing the 24 study regions into 30-meter cells. Each of the cells was classified as having a low, medium, or high probability of containing an archeological site. Areas that had not been adequately surveyed, and therefore lacked sufficient archaeological data to model accurately, were classified as "unknown."

The results giving "unknown" classifications have helped Mn/DOT determine where surveys are needed. The sensitivity results have been used to suggest project alignments or modifications that reduce the potential for impacts on cultural resources. This information has allowed Mn/DOT to expedite project clearance, reduce costs, and do a better job of protecting cultural resources.

Environmental Stewardship and Streamlining

With Mn/Model, fewer site surveys are necessary, saving Mn/DOT time and money. When surveys are necessary, they are more targeted and quicker to complete. In fact, within two years, the Mn/Model repaid its investment with survey and mitigation costs savings alone. Cost savings over the first four years of Mn/Model use have reached $3 million per year.

In addition, MnDOT staff estimate that Mn/Model allows Mn/DOT's Cultural Resources staff to clear more projects per year and that Mn/Model improves project turnaround time. In fact, some projects have saved one or two construction seasons in survey time alone.

Next Steps

There is now a movement toward integration among State agencies in Minnesota. Mn/DOT is working on an initiative to develop a common GIS database with the SHPO and the Minnesota Office of the State Archaeologist (OSA), and to enhance the OSA website. Currently, the site allows users to search for information on the location of burial sites. With

GIS for Environmental Stewardship and Streamlining
An Overview of State DOT Practices

extension of the site, all archaeological sites within the State would be searchable and able to be updated.

Texas

The partnership formed by the Texas Department of Transportation (TxDOT) and the Environmental Protection Agency (EPA) Region 6 to apply the GIS Screening Tool (GISST) to the NEPA process on the I-69 project exemplifies the potential of GIS to perform sophisticated analyses. Developed by EPA, GISST is a system that imposes a scoring structure on GIS coverages to inform decisionmaking and prioritize environmental protection. The system has many applications; however FHWA and TxDOT are using GISST as a screening tool for the NEPA Tier 1 Corridor Level decision. TxDOT uses the system to identify areas to avoid and to enable TxDOT decisions about where to concentrate resources for further studies at NEPA Tier 2. GISST has been designed to better understand the potential significance of single and cumulative impacts and to facilitate communication of technical and regulatory data with industry, the public, and other stakeholders.

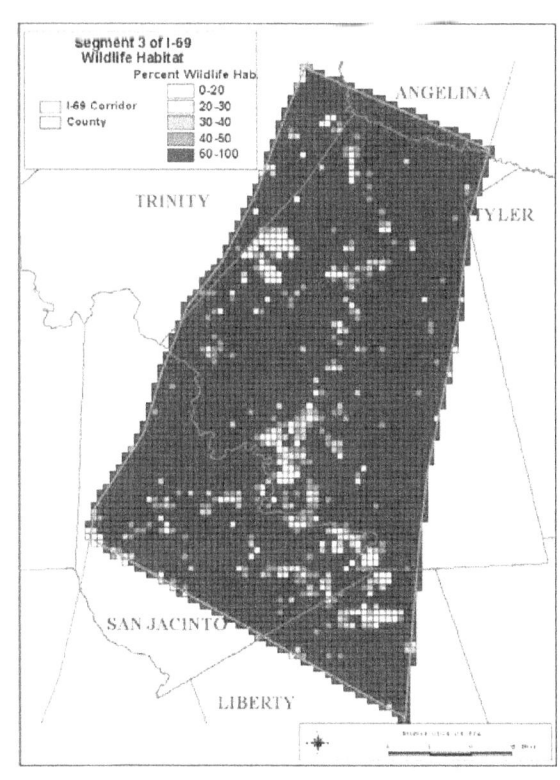

GISST is just part of the continuum of efforts TxDOT has undertaken related to I-69. While GISST has also been used for large scale screening of the corridor, the State and consultants are using the same environmental databases and QUANTM to define specific alignments for further environmental study within the corridor.

Another unique element of the I-69 alignment selection process is the partnership between TxDOT, the Texas Parks and Wildlife Department, and NatureServe,[10] a non-profit formed by the Nature Conservancy and others to collect and manage data about the status and distribution of species and ecosystems of conservation concern. Texas State law protects private property by prohibiting State agencies from releasing any information related to resources located on private land without the owner's consent. To get better quality data on endangered species, TxDOT partnered with the Texas Parks and Wildlife Department and NatureServe. These organizations produce Natural Heritage databases, which contain natural resource data for all lands within the State and provide an important data verification measure in lieu of public disclosure.

Environmental Stewardship and Streamlining

GISST is a valuable GIS4EST tool in several ways. TxDOT anticipates significant timesavings in the NEPA Tier 1 process as a result of using GISST. By explicitly establishing

---

[10] For more information see http://www.natureserve.org/.

GIS for Environmental Stewardship and Streamlining
An Overview of State DOT Practices

a clear rating system for environmental resources, the GISST makes the NEPA process more objective. This is largely because GISST helps to identify large-scale areas through its screening capabilities. The relatively quick and easy screening process offered through GISST points out 'red flags' to avoid and where additional information and analysis is needed at the NEPA Tier 2 stage.

In addition, FHWA and TxDOT officials also credit the GISST with an increase in trust and cooperation between historically disparate agencies, such as the EPA, the US Fish and Wildlife Service, the US Army Corps of Engineers, the Texas Parks and Wildlife Department, and the Texas Commission on Environmental Quality. Gaining consensus on the data and the criteria by which to rank features results in less conflict and more credibility for the transportation planning process. GISST encourages consistency by applying the same process to various decisionmaking points.

Of key importance for FHWA and TxDOT is that EPA endorsed the use of the GISST on the I-69 project. An important consideration that DOT officials must give to the development of GIS screening tools or models is whether or not EPA and other resource agencies will concur that the data is sufficient for the decision at hand. Consulting with the resource agencies prior to and during GIS development assures the DOT that its analyses will be accepted.

RE-ENGINEERING DOT BUSINESS PRACTICES

Virginia

After several years of developing in-house GIS capabilities, VDOT is leading GIS efforts in Virginia to catalogue transportation and natural resource data for use in transportation geospatial applications. VDOT officials expect that GIS will provide more than data management and map-making capabilities; they believe that GIS can change the business process within the DOT, fostering better communication and ultimately better decisionmaking. VDOT now boasts an Information Technology Application Division employing 120 people (State employees and consultants).

VDOT's GIS4EST work consists of several discrete projects. For instance, VDOT's Information Technology Application Division has assembled transportation and environmental data from internal DOT Divisions and resource agencies into one data repository: VDOT's Enterprise GIS. The Environmental Group previously stored

16

GIS for Environmental Stewardship and Streamlining
An Overview of State DOT Practices

transportation and natural resource data among 60-70 scattered databases and spreadsheets. These redundant systems represented a sizable waste of staff time and effort. With the Enterprise GIS, environmental staff can access spatial data at their desktops instead of searching through paper files or myriad, unintegrated systems.

VDOT also financially supported the creation of a Natural Heritage Resource Database, which was developed by the Virginia Natural Heritage Program (VNHP). The development of the natural heritage resource database has ensured that VDOT staff has easy access to reliable data essential to the NEPA process.

However, the GIS4EST application that represents a re-engineering of business processes is Comprehensive Environmental Data and Reporting (CEDAR), a spatially enabled project management tool. CEDAR, which VDOT initiated in the fall of 2002, is a workflow application with a spatial component that provides project management capabilities, a mechanism to track project progress, and a way to improve internal, interagency, and consultant communication. The project management capabilities of CEDAR enable users to notify users in other groups or agencies with questions and concerns, track projects, send email notification, and assign roles and responsibilities.

The first phase of CEDAR will provide a tool for project documentation and management for in-house users, as well as a way to track and monitor workflow. The first stage was culminated in a statewide training in the summer of 2004. Once security issues are resolved in the second phase, the Information Technology Application Division will implement web accessibility so that resource agencies and environmental consultants can also use the system. VDOT expects that providing access to resource agencies and consultants will greatly enhance communication in the NEPA process.

VDOT has developed and enabled CEDAR for use on all types environmental projects, including those that receive Federal funding and are required to be submitted to NEPA, as well as those that are fully funded by the State. While the latter projects are outside of the NEPA process, they are still required to undergo a State environmental review process that requires agency consultation.

Environmental Stewardship and Streamlining

While CEDAR is still in the development phase, VDOT is already reaping the benefits of the system on the Interstate-81 road-widening project, which will run the entire length of the State of Virginia. VDOT staff estimate that the geospatial data in CEDAR has enabled them to shave approximately 1,000 hours off the contract resulting in an estimated savings of $100,000. These savings have been realized because CEDAR has obviated the need for each consultant working on the project to go through the data collection and assimilation process. VDOT expects to see repeated savings through the use of CEDAR.

Florida

The Florida Department of Transportation (FDOT), along with FHWA, joined in a cooperative effort with Federal and State resource agencies to redesign the planning, permitting, and project review process. The resulting Efficient Transportation Decision Making (ETDM)[11]

---

[11] For additional information about ETDM, see http://etdmpub.fla-etat.org/.

GIS for Environmental Stewardship and Streamlining
An Overview of State DOT Practices

process has allowed for more efficient and effective incorporation of environmental data, project review, and technical assistance into transportation projects. By linking transportation, land use, and environmental resource planning, the ETDM has helped facilitate early and interactive involvement of all involved resource agencies and promoted the delivery of better and more consensus-based environmental outcomes.

Integral to the ETDM process is the use of GIS. FDOT has designed a GIS4EST application—the Environmental Screening Tool (EST)—allowing partnering agencies to share data electronically, compare analyses, and comment on proposed alternatives throughout the environmental review process. In long-range planning, agencies can evaluate cumulative impacts on a project and system-wide basis. The agencies are then able to consider the interrelationship between land use, ecosystem management, and mobility plans with an integrated approach. The EST represents a shift in how stakeholders collectively communicate, interact, plan, and manage transportation improvement projects. As a result, FDOT expects more efficient and effective environmental stewardship, along with reductions in delays, project changes, and challenges associated with project development, permitting, and consultation.

The University of Florida's GeoPlan Center[12] and its Florida Geographic Data Library (FGDL)[13] maintain the EST's geospatial data. EST features an Internet-based application linked to an electronic database system. Users can view and comment on the results of GIS analyses related to the environmental impacts and requirements of proposed project plans and alternatives through the EST interface. While previously, access to the EST was only available to members of an Environmental Technical Advisory Team (ETAT), groups formed specifically to complete the environmental review process, public access to EST went live in the Fall of 2004. The public can now review projects at the same time as agencies and submit their comments and concerns to their Community Liaison Coordinator.

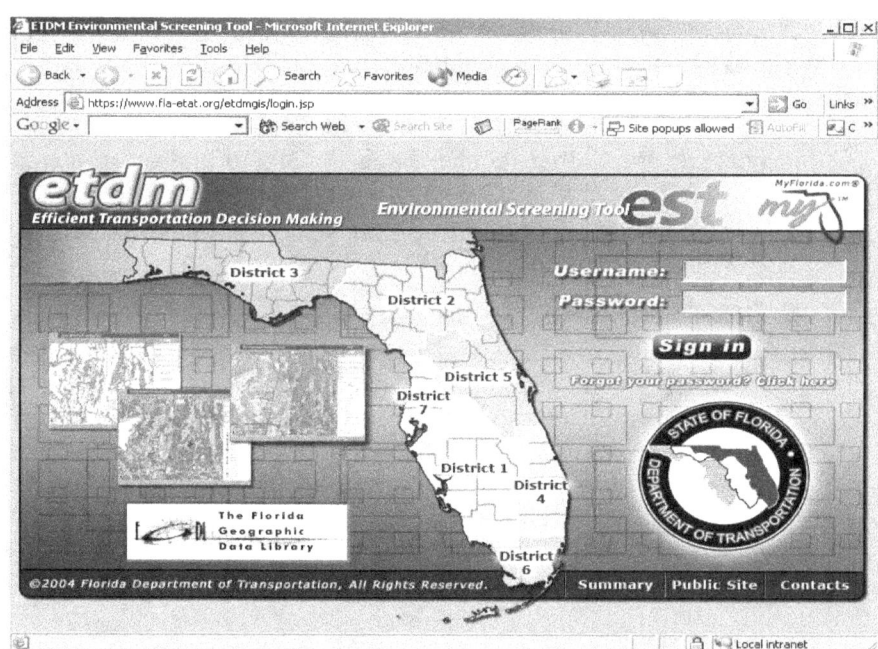

Each of Florida's seven districts has an ETAT, which consist of FDOT district staff and planning and regulatory staff from State resource agencies. During planning, ETATs can use EST in an advisory manner, providing input on regulatory and planning priorities. The ETATs can also comment on

---

[12] See http://www.geoplan.ufl.edu/project.html

[13] See http://www.fgdl.org/.

GIS for Environmental Stewardship and Streamlining
An Overview of State DOT Practices

avoidance, minimization, and mitigation options, allowing for a more accurate estimation of project costs. During project development, the role of the ETAT changes from advisor to coordinator. The ETATs use EST as a way to provide their input and technical assistance related to permitting decisions.

Environmental Stewardship and Streamlining

Through ETDM and EST, FDOT anticipates a more effective and timely decisionmaking process that does not compromise environmental quality. Since the project development phase will be incorporated into NEPA with the EST, FDOT estimates that the time required to complete the environmental review process will be reduced on the order of several years. The collaborative nature of EST may further enhance an expedited environmental review process. The Tool allows FDOT and collaborating agencies to visualize and address potential project flaws, while determining ways in which goals might be accomplished. EST promotes communication, concurrence, and early buy-in from all involved agencies—all crucial elements to speeding project implementation.

Another way in which EST is expected to streamline the environmental review process is by flagging and resolving concerns early in the process. FTATs will be able to focus on key environmental issues in their districts and will be better prepared to convey these issues to each other. The interagency communication and detailed reviews that EST supports should help ensure that ETAT concerns are noted and that any project disputes can be resolved before funding. In one case, a project was even removed from an MPO's long-range plan based on the concerns submitted by the ETAT, thus saving labor hours and project funds on an unworkable project.

EST allows early identification of avoidance and minimization options and allows partnering agencies and the public to express their concerns early on—allowing better integration of agency and community perspectives. FDOT expects that this will reduce challenges and litigation at late stages in the process

Next Steps

Performance measures for achievement of overall time frames are being created though obtaining agreement on a proper measure of success has been difficult. FDOT has created a task force, which has a draft performance white paper currently under review.

GIS4EST CATALYSTS

In the previous cases, the catalyst for GIS4EST development varied according to local conditions. For some States, the catalyst was a large project requiring the collection, systemization, and management of a large amount of data. In fact, for three of the following cases (Indiana, Arkansas, and Texas) this project was the I-69 NAFTA highway.

For other States, the catalysts were related to data needs. For example, in Ohio and Minnesota, State DOTs entered into partnerships with SHPOs to develop spatial databases of cultural resources. In Washington, the State DOT developed a GIS application to better catalogue resource data and provide a forum for interagency input.

GIS for Environmental Stewardship and Streamlining
An Overview of State DOT Practices

In other cases, the purpose for the development of a GIS4EST was explicitly to promote environmental stewardship and streamlining. In Florida, for example, the development of the ETDM and EST was intended to support a broader effort to re-engineer the environmental review process itself. VDOT's CEDAR was also designed to reshape the way the entire environmental process is conducted.

These cases illustrate that there are many approaches in launching a GIS4EST.

LESSONS LEARNED AND RECOMMENDATIONS

While GIS4EST development and experience varies among States, the eight cases reveal five crosscutting lessons:

- Look Beyond Initial Costs
- Start Small and Build as You Go
- Collaborate and Build Partnerships
- Develop Applications Tailored to Local Needs
- Focus on Evaluations Measures Before You Start

LOOK BEYOND INITIAL COSTS

The up-front costs required for development, implementation, and maintenance of any GIS are not trivial. Acquiring the necessary hardware and software, hiring expert GIS staff, and training current staff to use the resulting systems are significant expenses. However, many systems achieve full cost recovery within the first several years. It is important for State DOTs, therefore, to make a long-term commitment to GIS to realize its full benefits. Sources of long-term cost savings include:

- *Reducing redundancy in data collection and maintenance systems.* GIS4EST can lessen and eventually eliminate the need to collect—or pay consultants to collect—the same data for each new project. Several DOTs reported that prior to the development of the GIS4EST, data was scattered across the agency in redundant and unintegrated systems.

- *Better relationships with resource agencies and the public.* Reaching agreement with resource agencies on the spatial distribution of natural and cultural resources can minimize misunderstandings and confusion that often delay the environmental review process and result in cost overruns. In addition, because GIS4EST can be a mechanism for the collection and management of interagency and public input, State DOTs can use these systems to collect more meaningful input from these sources. This can result in State DOTs having a better understanding of potential flash points in a project and moving to resolve them more quickly.

- *Minimizing the number of field surveys.* Upon obtaining EPA or other resource agency buy-in, predictive modeling and screening tools are successful ways to reduce the number of costly field surveys while ensuring environmental protection.

GIS for Environmental Stewardship and Streamlining
An Overview of State DOT Practices

- *Integration of planning and NEPA processes.* Through easy visualization of resource distribution, transportation agencies can develop plans that avoid impacts to the most sensitive or critical resources. One of the clearest benefits of GIS4EST is the systemization and categorization natural and cultural resource distribution. The first step to avoiding and minimizing impacts to these resources is knowing where they are on the landscape. In this way, conflicts can be resolved and concerns addressed in the planning stage before significant resources have been invested in unworkable alternatives.

For example, FDOT EST's implementation costs were not inexpensive. Before FHWA provided funding for the development of the EST, FDOT had invested roughly $400,000. Since that time, approximately one million dollars in State funds and one million dollars in Federal funds have been directed towards the expansion of the EST and FDOT's GIS program. Despite these upfront costs, FDOT expects that the EST will begin to pay for itself through future cost- and time-hour savings during NEPA review. Mindful of the stewardship commitment in GIS4EST, FDOT officials also anticipate that the EST will help create savings beyond measure by preserving the environment.

Furthermore, training non-GIS staff to use GIS4EST applications requires both the development of training programs as well as the staff time to run the sessions. However, these investments can yield high returns. In Washington, State DOT officials found that investing the time up-front to train staff in simple mapping techniques saves staff time by reducing the need to "hunt down" necessary data. In the case of the Workbench in Washington State and the EST in Florida, the application also conserved the time of the GIS staff who, freed of obligations to create simple maps for staff, can now devote more time to assuring data quality and seeking out new data sources.

START SMALL AND BUILD AS YOU GO

Starting small can allow GIS4EST development to proceed in manageable way. Several States in various stages of development have started with small projects and used those experiences to demonstrate results and acquire feedback, as well as build staff expertise. State DOTs can use small projects to demonstrate the value of GIS to internal staff, resource agencies, consultants, and the public before undertaking more ambitious projects.

For INDOT and its consultants, the GIS for Southwest for Indiana was a logical way to systemize and store the enormous amount of data generated from the environmental impact assessment of I-69. The development of this GIS also served as a way for INDOT to get its feet wet with GIS development and foster the development of key relationships, such as with the IGS. Proceeding slowly afforded INDOT the opportunity to refine this regional GIS into a statewide system. Once the GIS for Southwest Indiana had demonstrated its usefulness, INDOT and its partners were more invested in the development of a larger system.

Clear project phasing can be crucial in managing GIS growth. For example, in Ohio the development of the GIS was divided into two phases. First ODOT entered into an agreement with OSHPO to assist in the development of a comprehensive electronic spatial database of the OHI and OAI. Then OSHPO presented the primary results of the system to ODOT. Both agencies approved of the work and agreed on the potential utility of the system particularly in the early planning stages. The clear benefit and utility of the initial investment in the GIS project generated more enthusiasm for the second phase, which enabled ODOT to access a major portion of the OSHPO data files electronically. Once OSHPO demonstrated results,

GIS for Environmental Stewardship and Streamlining
An Overview of State DOT Practices

the agency—with ODOT's financing and input—developed the second phase of the project. Dividing the development of the system into these two phases allowed OSHPO to recruit and develop staff expertise in the system and also ensured a period of reflection and strategizing before launching the second phase of the project.

Starting small can also help you design for change—or build in the flexibility you need into your system as you proceed gradually. Moving too fast to develop complex GIS4EST applications may later lead to considerable effort to retrofit the application for unforeseen uses. For example, WSDOT officials recommend that if States are thinking of developing an application for a narrow use—such as project scoping—that they get a sense of how other users might also use a GIS. Designing the GIS with as thorough a sense as possible of future uses would obviate the need to retrofit the application later.

COLLABORATE AND BUILD PARTNERSHIPS

In many States, GIS4EST projects are already off the ground as agencies put this technology to work on different projects. In these cases, DOT officials should build on what has already been done. TxDOT's experience illustrates this well. The GISST was an EPA-developed GIS application already in existence when the I-69 corridor was proposed. Instead of developing their own GIS, TxDOT officials partnered with EPA to use the existing GISST framework as a screening tool for the I-69 project. An additional benefit of using EPA's screening tool is that TxDOT was assured of EPA's buy-in on the results of the analysis from the outset.

In Arkansas, officials noted that related agencies should be brought "on-board" early in the process and that it can be challenging to get other agencies up to speed on GIS use, especially if in the past the technology has been viewed as being "a little different." Early involvement and communication can lead to agency buy-in.

The appropriate path to take depends on the pattern of interagency GIS development in a State. If State agencies are working together to develop a statewide consortium, it may be easy to contribute to and benefit from that effort. However, if that process is not producing results in the near-term, acting independently to develop a GIS may yield positive results. The time and cost savings that can be achieved by streamlining the environmental review process alone may enable the transportation department to achieve cost recovery for the GIS, even if other agencies do not contribute financially to it. When you are called upon to be a trailblazer, it is important to identify GIS champions that can help build support among various agencies for GIS technologies. For example, in Virginia having a champion in the Transportation Commissioner helped develop support for GIS among all levels of the Department. In Indiana, the GIS Atlas for Indiana may prove to be the catalyst in the development of a statewide GIS consortium.

In addition, if your agency is called upon to be a trailblazer in developing a GIS4EST, you will want to consider strategic partnerships as soon as possible. One way might be through the creation of a GIS Technical Committee such as in Minnesota or a statewide GIS consortium such as in Washington State.

In any case, it is important for State DOTs to consider the interoperability of your GIS4EST with other agencies' systems. For instance in Ohio, a major initial stumbling block for ODOT was whether to use the GIS software that SHPO staff was accustomed to using or whether the system should be based upon ODOT's software. By developing a web portal interface

GIS for Environmental Stewardship and Streamlining
An Overview of State DOT Practices

between the two, the staff at both agencies was able to continue to do their work in the GIS environment in which they are most comfortable.

Lastly, State DOTs should not overlook local colleges and universities as potential partners in the development of a GIS4EST. The training and technical assistance that these partners can provide can be a boon to GIS4EST development. In Indiana, INDOT and its consultants partnered with IGS to actually collect and maintain the spatial database. In addition, the University of Florida's GeoPlan Center and FGDL play a key role in maintaining FDOT's EST.

GIS4EST APPLICATIONS TO MEET LOCAL STREAMLINING CHALLENGES

Because GIS4EST applications can vary from screening tools to re-engineering the entire environmental review process, it is important for State DOTs to consider their priorities in environmental stewardship and streamlining. What are the most contentious areas of NEPA where data enhancements and partnerships with agencies can achieve greatest results? Once these issues are determined, the GIS4EST application should be tailored to meet those needs.

For some States, the greatest potential for promoting environmental stewardship and streamlining may lie in a screening tool. By identifying critical resources along a transportation corridor that warrant focused investigation, GIS technologies can save DOTs time and money by focusing detailed analyses only on the issues and areas that truly warrant them. For other States, evaluating cumulative impacts (such as with GISST) or predicting likely locations of archeological resources are priorities. Utilizing GIS to catalogue and assess cumulative impacts over time to a given resource can greatly inform the decisionmaking process and reduce workload and stress to agencies that struggle with this complex issue.

FOCUS ON EVALUATION MEASURES FROM THE OUTSET

Cost benefit analyses are crucial evaluation tools to build continued support for your GIS4EST. However, these analyses are complicated if State DOTs do not consider evaluation measures from the outset. Once a GIS4EST application has been finalized and *before development begins*, State DOTs should question what aspect of environmental stewardship and streamlining it will address and how progress to that goal will be measured.

WSDOT staff noted the importance of collecting baseline information about the tasks that a given GIS4EST application seeks to improve. In the case of the Environmental GIS Workbench, WSDOT wishes that they had data on how long it took staff to complete scoping work before the application was developed. This baseline information would make possible a more robust evaluation of the Workbench.

For example, a VDOT official noted the importance of developing evaluation measures because this information is essential to justifying GIS4EST applications and ensuring long-term funding. To evaluate these projects, this official recommends documenting consultant fees as a fraction of the total project cost. Apply these factors to the number of hours of consultant time to develop an estimate of time and money saved. While future savings may be difficult to estimate once consultants stop including data collection and assimilation time into the scope of work, this may merely be a positive sign that true cost savings are being built into the process.

GIS for Environmental Stewardship and Streamlining
An Overview of State DOT Practices

APPENDIX A: INTERVIEW GUIDE

1. What is the application exactly and how was it developed?
2. Were there particular local circumstances that created the need for this GIS application? (i.e. a contentious local planning process, political considerations, unusually environmentally sensitive land, etc.)
3. When did you realize that there was a need for a GIS application? What was this application a response to? What need does it fill?

    <u>Follow-up points</u>: *Data redundancy? Need to comply with national standards? A lack of technical capability at agency level?*

4. How expensive has the project been? Who is paying for it?
5. Where has the data come from?
6. Who manages the data? (hosts it, standardizes the data and metadata, updates the database)
7. How is staff/the public being trained in using this application?
8. Is it an internal review system/ multi agency coordination tools or is there a general public version?
9. Is the application one that will be used to track continuing / long term changes or is it just related to a specific project?
10. Has the GIS application fostered new types of inter-agency collaboration? Can you estimate how it has improved interagency cooperation? (only if it somehow hasn't come up earlier)
11. Can you estimate how much it sped up project implementation?
12. Are there environmental issues that we can now address that couldn't before?
13. Are there other ways this application is helping you make better environmental decisions or streamline the process?

    <u>Follow-up Question</u>: *How useful has the spatial data that you've collected been for NEPA practitioners? How do they use it in the NEPA process? At what stage of the NEPA process is this application useful?*

14. Did the application save the State money / staff time? How many man-hours have been saved? How much has developing the application cost?
15. What feedback have you gotten from the public/partnering agencies/whoever-else-is-involved about this GIS application?
16. What have you learned from this project? Biggest successes? Biggest obstacles?
17. What words of advice would you give to other States that are thinking of undertaking a similar project?

GIS for Environmental Stewardship and Streamlining
An Overview of State DOT Practices

APPENDIX B: INDIANA CASE STUDY

PRACTICE TITLE: *GIS for Southwest Indiana, GIS Atlas for Indiana*
CONTACT: *Janice Osadczuk*
EMAIL: *josadczuk@indot.state.in.us*
PHONE: *(317) 232-5468*

Introduction and Background

The Indiana Department of Transportation (INDOT) has compiled over 170 layers of information commonly used for planning and environmental purposes to create a statewide Geographic Information System (GIS). Entitled A GIS Atlas for Indiana, this statewide system was designed to foster the consideration of potential environmental impacts early in the project development process. The main features of the GIS Atlas for Indiana website include an Interactive Mapping Server (IMS) enabling the construction of maps via the Internet; downloadable files of spatial data, including reference, demographic, infrastructure, environmental hydrologic, and geologic data; and metadata text files for each data layer. Many of these layers are available to the public; some more sensitive layers, such as karst topography, have restricted access. In addition to the environmental and transportation planning applications of these data, this spatial information can also aid rural and urban planning and business development.

The project had its inception in 1999 when INDOT began the environmental review process for the Interstate 69 (I-69) project in southwestern Indiana, a roadway that will span 142 miles in Indiana and a study area of 26 counties.[14] I-69 is also known as the "NAFTA highway"—a congressionally mandated, 1,600-mile interstate highway stretching from the Mexican border in Brownsville and McAllen, Texas to the US-Canadian border in Detroit, Michigan. I-69 was chosen in 2002 as a streamlining pilot project[15] under TEA-21, Section 1309.

INDOT, with the assistance of an engineering and environmental consulting firm, utilized a tiered environmental document for the project. Recognizing since they were going to collect environmental data over such a large area, INDOT sought a way to permanently store this work for future use. INDOT identified the development of a GIS database as the best way to accomplish this goal. INDOT and its consultants, therefore, subcontracted with the Indiana Geological Survey (IGS) of Indiana University to compile the GIS for Southwest Indiana,[16] which now contains 173 layers of geospatial information.

The GIS for Southwest Indiana was intended to be a project-specific GIS for the I-69 corridor. However upon its completion, INDOT could clearly see the value of expanding its scope to the entire State, and in April 2002 INDOT granted IGS funds to expand the GIS into a statewide computer-based atlas. Entitled the GIS Atlas for Indiana,[17] this statewide project

---

[14] The Final EIS was released on the project in December of 2003 and is available at http://www.in.gov/dot/projects/I69/.

[15] For a complete list of the priority projects, go to http://www.fhwa.dot.gov/stewardshipeo/pplist.htm.

[16] Available at http://igs.indiana.edu/arcims/southwest/viewer.htm

[17] Available at http://igs.indiana.edu/arcims/statewide/index.html

GIS for Environmental Stewardship and Streamlining
An Overview of State DOT Practices

now contains 206 layers of free downloadable spatial data. IGS maintains the two projects—the GIS for Southwest Indiana and the GIS Atlas for Indiana—as separate IMSs on their website, as well as two additional IMSs not sponsored by INDOT (a Lake Rim GIS IMS and an Indiana Coal Mine Information Systems IMS).

Challenges

The GIS for Southwest Indiana—and later the GIS Atlas for Indiana—emerged as a practical way to store the extensive data accumulated for one complex project, rather than as a solution to a particular issue. While several State agencies in Indiana earlier had expressed a desire to create a statewide clearinghouse for geospatial data, little progress was being made to move the idea forward. When the I-69 project presented the opportunity for the development of a GIS, INDOT made the decision to advance the statewide GIS project on their own, without the financial and staffing resources of other agencies in the State. Fortunately, the benefits of the GIS for Southwest Indiana and the GIS Atlas for Indiana for INDOT were great enough that they far outweighed the costs of developing it, even though INDOT was the only agency to dedicate funds to the project.

As a consequence of INDOT's decision to initiate the development of a statewide GIS, INDOT now has demonstrated expertise in GIS and is now quickly setting the standard for GIS development in Indiana, defining both the "look" of the website, as well as the format and distribution of spatial data. The Indiana Department of Environmental Management (IDEM), the Federal Emergency Management Agency, Natural Resource and Conservation Service, US Geologic Survey, as well as several universities have all contacted INDOT and its consultants requesting information about these GIS services, and IDEM has asked them to add more of their data to the site.

The Details

Collecting Spatial Data - IGS researchers work with INDOT's consultants to obtain data from Federal, State, and local agencies and other sources. The information is then edited and processed into a standardized format. The project partners are asking the public and other agencies what other layers they would like to see incorporated into the GIS. The IGS site is not utilized to show layers that provide specific alternatives for various INDOT projects because these layers would change so frequently that they would be out of date almost as soon as they were put online.

INDOT's relationship with IGS has been mutually beneficial. IGS has benefited from the positive publicity from hosting the GIS website; in 2003 the IGS website had over 2.5 million hits. In addition, INDOT has been able to build a more robust GIS by leveraging the expertise and existing resources of IGS. Much of the geospatial data that are a part of the GIS for Southwest Indiana and the GIS for Indiana have originated from IGS.

Use of the Statewide GIS - One use of the GIS is unique to the topography of Indiana. Indiana's karst topography, particularly in the southeastern part of the State, gives the area its many caves and sinkholes. Cave features, therefore, are a salient feature in transportation planning in Indiana, and the GIS Atlas for Indiana has enabled better planning of alternative alignments that minimize impacts on these features. While geospatial data of karst topography are not available to the general public for security reasons, IGS and INDOT make it available to contractors working on specific projects.

GIS for Environmental Stewardship and Streamlining

An Overview of State DOT Practices

Funding the GIS - The cost of developing the GIS for Southwest Indiana (both the database and the web-based access) was about $100,000, a fraction of the cost of reaching a typical Record of Decision (ROD) which INDOT officials estimate usually cost around $500,000. INDOT funded the entire GIS for Southwest Indiana project through the environmental review budget for the I-69 project. When INDOT subsequently decided to expand the project to the entire State, they applied for and received a State Planning and Research (SPR) grant of $850,000 (over 3 years). INDOT is matching 20 percent of the grant. INDOT is not certain of the long-term costs but expects that it will be approximately $10,000 per year, which includes keeping existing layers current as well as adding additional layers yearly.

Stewardship and Streamlining

INDOT and its consultants use GIS on most NEPA documents, whether the project is several hundred feet or over a hundred miles. INDOT recognizes GIS as a useful way to get an early sense of environmental features existing along a proposed alignment and devise ways to minimize impacts to those resources from the outset. INDOT's consultants note, "the number one principle of mitigation is to avoid bad projects," and GIS has been particularly useful because it helps them visualize potential impacts early. An example of this type of early visualization is that INDOT's consultants provide agency partners with paper copies of maps derived from the Statewide GIS prior to bus trips to survey the project area. According to INDOT and its consultants, the use of the maps "triggers a whole thought process and enables them to do planning better before they leave the office."

The GIS Atlas for Indiana also minimizes the need for information to be "chased down" and reduces project costs and time. Sensitive resources can be avoided early on, when the greatest flexibility in terms of avoiding impacts exists. INDOT's consultants note that whenever a firm works on an EIS using the GIS, they have better quality data, can get to "common ground" more quickly, and save time and money. Having a commonly referenced set of data that is up-to-date has both reduced inter-agency and public conflict and promoted better environmental decisions. The contractors working on the I-69 project also cite the GIS as particularly useful in building credibility with the public on EISs. Providing the public with the same data that are being used to evaluate alternative alignments minimizes unnecessary conflict over "what's out there" and helps build consensus on a Locally Preferred Alternative more quickly.

INDOT expects to quickly recover their investment in the GIS and estimates that the time and cost savings achieved through only two or three environmental reviews will pay for the system. While INDOT notes their preference to partner with other organizations to develop the GIS into a consortium, INDOT is also willing to shoulder future maintenance costs themselves, if necessary, because the cost savings to the transportation agency are so substantial.

Next Steps and Lessons Learned

Starting small can allow the GIS to proceed in manageable way, allowing partners to develop expertise gradually and clarify the purpose of the GIS over time. For INDOT and its consultants, the GIS for Southwest for Indiana was a logical way to systemize and store the enormous amount of data generated from the environmental impact assessment of I-69. The

GIS for Environmental Stewardship and Streamlining
An Overview of State DOT Practices

development of this GIS also served as a way for INDOT to "get its feet wet" with GIS development, foster the development of key relationships, such as with the IGS, and afforded INDOT the opportunity to refine this regional GIS into plans for a statewide system. Once the GIS for Southwest Indiana had demonstrated its usefulness, INDOT and its partners were more invested in the development of a larger system.

Consider strategic partnerships. According to INDOT and its consultants, agencies should "create as many relationships as you can because you cannot do it without them." These partnerships can be very useful, especially in hosting and maintaining a website, if this is a touchy issue. Universities, such as Indiana State University, are frequently excellent partners and benefit from being somewhat removed from the political process.

Be a GIS trailblazer. In Indiana, the I-69 project presented a prime opportunity to use and save the environmental data collected during the NEPA process for a larger GIS project. However, at the same time, an interagency process to develop a statewide GIS consortium was stalling. INDOT made the decision to use the collection of these data to create a small GIS for the project area, A GIS for Southwest Indiana. Once INDOT and others were convinced of the usefulness of this GIS, it expanded into the statewide project, A GIS Atlas for Indiana. The GIS Atlas for Indiana may prove to be the catalyst in the development of a statewide GIS consortium.

INDOT's experience in developing a GIS may be instructive to State DOTs. If State agencies are working together to develop a statewide consortium, it may be easy to contribute to and benefit from that effort. However, if that process is not producing results in the near-term, acting independently to develop a GIS may yield positive results. The time and cost savings that can be achieved by streamlining the environmental review process alone may enable the transportation department to achieve cost recovery of the GIS, even if other agencies do not contribute financially to it.

GIS for Environmental Stewardship and Streamlining

An Overview of State DOT Practices

APPENDIX C: ARKANSAS CASE STUDY

PRACTICE TITLE: *GIS for the I-69 Southeast Arkansas Connector*
CONTACT: *Randal Looney*
EMAIL: *Randal.Looney@fhwa.dot.gov*
PHONE: *(501) 324-6430*

Introduction and Background

Transportation professionals in Arkansas are currently constructing a stretch of highway that will ultimately connect Interstate 69 (I-69) to the existing Arkansas interstate highway system. The roadway, known as the Southeast Arkansas Connector (I-69 SE-Connector), is the first of three I-69 projects in Arkansas. The purpose of each of the projects is to improve traffic flow and safety, while enhancing capacity. In order to accomplish these goals efficiently, the Arkansas State Highway and Transportation Department (AHTD) applied new technologies and means of environmental analysis during stages of project planning and development.

AHTD utilized Geographic Information Systems (GIS) technology to streamline the transportation decisionmaking and permitting process for the I-69 SE-Connector. GIS was viewed as a way to share and consolidate environmental and engineering data. The technology provided AHTD a tool with which large amounts of study area information could be refined and analyzed, allowing for the efficient screening of project alternatives.

However, one of AHTD's earliest GIS efforts was digitizing General Land Office (GLO) survey maps from the 18th and 19th centuries. These maps reveal features that have since disappeared from the visible landscape—such as historic streambeds, Native American mound sites, and homesteads. Knowing where these unique features are helps transportation planners avoid or minimize impact to these features.

The Challenge

The development of the I-69 SE-Connector in Arkansas encompassed three projects and involved multiple Federal, State, and local agencies. The project, like other transportation system improvements, was also governed by numerous environmental regulations. AHTD was faced with developing an effective manner by which to foster early coordination with resource agencies, the public, and Native American tribes while efficiently addressing the requirements of the National Environmental Policy Act (NEPA) review process.

The Details

Before the I-69 SE-Connector project began, AHTD had been exploring innovative ways in which GIS technology could be applied. In the early 1990s, a consulting firm signed a contract with the AHTD to perform GIS work. Initially, the consulting firm was hired in order to provide AHTD with clear, comprehensive maps of proposed project alignments. Due to the success of the map creation and overall GIS integration and implementation, AHTD began to understand the extent to which GIS could aid with the completion of other project tasks. Specifically, AHTD believed that GIS, along with consultant expertise, could help determine project impacts for Environmental Impact Statements (EIS) in a more efficient manner. In AHTD's view, the technology could provide a quick, accurate, and precise instrument for the generation of maps detailing a project's environmental constraints.

GIS for Environmental Stewardship and Streamlining
An Overview of State DOT Practices

During the planning and development of the I-69 SE Connector, the use of GIS for environmental constraint mapping came to the forefront as a potential streamlining tool. In the project, two-mile wide preferred corridors, each with 300-foot alignments, were delineated. GIS coverages containing environmental constraint data were overlaid on each of the preferred corridors. This allowed for quick but thorough identification of Draft EIS alternatives. The maps and analyses that GIS investigation provided also gave partnering agencies and communities tangible examples of how various project alternatives would impact environmental, cultural, and economic resources. Partnering agencies supported GIS use because project steps occurred more quickly. The public especially welcomed GIS use and appreciated the map visualizations. AHTD noted that public participants were eager to learn how the project would affect their neighborhoods, properties, and houses. By providing this information, AHTD was able to garner and evaluate more easily public response to the subtle differences of proposed alternatives.

Initial costs of GIS program implementation were roughly $100,000, and currently, AHTD maintains a small GIS unit with a staff of five. However, three years ago the GIS staff was two. AHTD expects the in-house GIS unit to continue its present growth. As GIS benefits are realized through new projects, increased funding is anticipated.

Stewardship and Streamlining

GIS expedited NEPA project development of I-69 Connector. Previously, an average of 62 months was required to move a project from the Notice of Intent (NOI) to the Record of Decision (ROD) stages of project development. For the 1-69 SE Connector project, the ROD was signed in 26 months, a 58 percent reduction in time. Though the expedited schedule was due in part to efforts to coordinate with other agencies early and often, AHTD attributes a majority of the timesavings to the use of GIS.

The environmental review process proceeded so quickly that AHTD began to believe the project might have been overstepping the purpose of NEPA. The Department drafted a letter to FHWA Headquarters requesting guidance and approval of the appropriateness of applying GIS to the I-69 SE Connector (and future) project(s). In the response, FHWA concurred with AHTD's use of GIS, noting that GIS is a complement to—not a substitute for—public involvement, scoping, and alternative development and analysis.

Lessons Learned

AHTD learned several lessons regarding the implementation of GIS from the I-69 SE Connector project. The AHTD insight acquired through experience may be beneficial to other transportation agencies exploring GIS use for environmental streamlining objectives. Lessons learned include:

Obtain Interagency Buy-in. Related agencies should be brought on-board early in the process. It can be challenging to get other agencies up to speed on GIS use, especially if in the past the technology has been viewed as being "a little different." Early involvement and communication can help interagency GIS buy-in gain momentum.

Look Beyond Upfront Costs. Initial costs of implementing GIS initiatives and strategies will likely be offset in the long run by total project savings. AHTD has noted that one data layer can make a difference in a project, or an entire program, and it would be a mistake to not invest in additional layers.

GIS for Environmental Stewardship and Streamlining
An Overview of State DOT Practices

Ask for and Secure Cooperation and Assistance. AHTD found that related agencies were very willing to share money and other resources, including technical assistance. FHWA Arkansas Division, the Forest Service, the United States Army Corps of Engineers, the United States Geologic Survey (USGS), and National Resources Conservation Service (NRCS) are agencies that helped AHTD implement a GIS program.

Share Data Layers. Easily accessible data layers can facilitate streamlining efforts. Readily available environmental data layers allowed AHTD to conduct GIS analyses earlier in the environmental review process. Some organizations that AHTD has shared data with include the Arkansas Land Commission and the University of Arkansas Fayetteville Center for Advanced Spatial Technology (CAST).

Next Steps

Currently, AHTD is evaluating other ways to use GIS to streamline the NEPA process for projects requiring an EIS. The DOT is conducting an archaeological survey to compile all of Arkansas' archaeological data in a web-based GIS format. The GIS will be accessible to archaeologists and transportation professionals on an intranet system. AHTD expects the system to aid in tribal consultation. Similarly, AHTD is also currently developing a historic bridge management GIS.

AHTD and the Federal Highway Administration (FHWA) Arkansas Division hope to expand their use of GIS by participating in the Federal GIS Users Group, a consortium of Federal agencies within Arkansas that collaborates to share geospatial data. Group meetings present a forum for agencies to share GIS data, discuss projects that may affect other participating agencies, and reduce duplications in data gathering. Recently, an AHTD staff member gave a GIS Workshop in northwestern Arkansas for the GIS Users Group. AHTD looks forward to developing additional workshops and training courses as staff time and funding allows.

GIS for Environmental Stewardship and Streamlining
An Overview of State DOT Practices

APPENDIX D: OHIO CASE STUDY

PRACTICE TITLE: *GIS Mapping Effort; Cultural Resources GIS*
CONTACT: *Tim Hill or Paul Graham*
EMAIL: *Tim.Hill@dot.state.oh.us; Paul.Graham@dot.state.oh.us*
PHONE: *(614) 644-0377 for Tim Hill, or (614) 466-5099 for Paul Graham*

Introduction and Background

Despite large potential gains toward aiding interagency cooperation and streamlining the environmental review process, financial constraints on resource agencies often hinder efforts to develop Geographic Information Systems (GIS) to catalogue resource distribution. In order to overcome these constraints, the Ohio Department of Transportation (ODOT) formed an innovative alliance with the Ohio Historical Society/Ohio State Historic Preservation Office (OHS/OSHPO). The resulting partnership led to the development of a GIS-based on Mapping and Preservation Information Technology (MAPIT)[18] software to document over 120,000 Ohio Historic Inventory (OHI) and Ohio Archaeological Inventory (OAI) features, such as individual properties and historic districts listed on the National Register of Historic Places in Ohio.[19]

ODOT's Office of Environmental Services (OES) was instrumental in initiating this streamlining process. The concept of the GIS emerged in the late 1990s when OES held meetings with OSHPO to determine what actions could expedite data collection to streamline the NEPA process. Because OSHPO needed additional funding to develop such a system, ODOT agreed to help finance this project. This joint development of the GIS was a win-win situation for both agencies; OSHPO obtained the resources they needed to thoroughly systematize knowledge of cultural resources the State, and ODOT gained electronic access to the newly formed cultural resource database at OSHPO. These data inform the way ODOT develops potential transportation alignments and are invaluable for planning purposes, as well as for the National Environmental Policy Act (NEPA) process, in general.

The Challenge

The primary challenge in developing the GIS was to create an interface between the different GIS software packages utilized by OSHPO (ESRI's ArcView) and ODOT (Intergraph's GeoMedia). Both agencies were reluctant to invest the time and financial resources necessary to convert to the other agency's GIS software. In fact, the original concept about how to develop this system (putting data on MAPIT web browser) was not workable because ODOT was disinclined to purchase and deploy a new system.

ODOT and OSHPO developed a solution to this problem: create a third Internet-based system to act as an intermediary between MAPIT and GeoMedia. The use of this

---

[18] *The Mapping and Preservation Inventory Tool* (MAPIT) is a National Park Service adaptation of ESRI's ArcView. MAPIT organizes historic resource inventories in a computerized database with sophisticated mapping capabilities. See http://www.conservationgis.org/links/historic.html or http://www.cr.nps.gov/map.htm.

[19] See the ESRI report at http://gis.esri.com/library/userconf/proc02/abstracts/a0809.html

GIS for Environmental Stewardship and Streamlining

An Overview of State DOT Practices

intermediary system obviates the need for either agency to purchase an entirely new GIS software package or for staff to invest time in learning a new system. With the current configuration, ODOT staff can use the GeoMedia software on their desktop computers to read OSHPO's ArcView data via the Internet portal.

The Details

The development of the GIS was divided into two phases. Once OSHPO demonstrated results, the agency—with ODOT's financing and input—developed the second phase of the project. Dividing the development of the system into these two phases allowed OSHPO to recruit and develop staff expertise in the system and also ensured a period of reflection and strategizing before launching the second phase of the project. Total cost of Phases I and II were $375,000. This funding came from State planning and research funds, as well as from FHWA contributions.

Project Phasing

Phase I: 1998 – 2000: In 1998, OES entered into an agreement with OSHPO to assist in the development of a comprehensive electronic spatial database of the OHI and OAI. In February of 2000, OSHPO presented the primary results of the system to ODOT. Both agencies approved of the work and agreed on the potential utility of the system particularly in the early planning stages. The clear benefit and utility of the initial investment in the GIS project generated more enthusiasm for the second phase.

Phase II: 2001 – 2003: In 2001, OSHPO began the second phase of the GIS project to enable OES to access a major portion of the OSHPO data files electronically. These data are now used to plot archaeological and historic site locations (including the National Register of Historic Places properties), consider preliminary alignments, and make early evaluations of potential environmental impacts in a transportation project area. Quick access to this information facilitates initial record searches and the development of a "model" for transportation projects. As mentioned above, ODOT and OSHPO overcame the challenges of integrating two different software packages by developing a web portal interface to import data from MAPIT into a GeoMedia working environment.

GIS Spin-off Projects:

ODOT receives quarterly reports from OSHPO with updated OAI and OHI data. These additional data have led to the expansion of the GIS. An example of this is the Predictive Modeling Program, which was developed as a result of data from previously surveyed project areas. The archaeological data that SHPO collects from all Section 106 surveys, which include transportation, Department of Housing and Urban Development, Department of Natural Resources, surface mining, and pipeline surveys, are collected into a database and integrated into the GIS. These new data are leading to the development of a Predictive Modeling Program, which will enable OSHPO to identify correlations between current and previous archaeological and natural features. ODOT expects that the Predictive Modeling Program will be a huge asset to the agency not only in identifying known archaeological resources but also in predicting the location of potentially significant cultural resources that might impact transportation planning and the development of alternatives.

GIS for Environmental Stewardship and Streamlining
An Overview of State DOT Practices

Another project that has emerged from the initial GIS project is a Drainage Areas Map, which delineates in digitized and color-coded form the drainage areas of all 30 major, as well as some minor, streams in Ohio. ODOT uses these data in consultation with Federally-recognized American Indian tribes.

Finally, ODOT is in the process of integrating spatial data of all properties that are eligible for the National Register of Historic Places. This information is important for transportation decisionmakers because if a historic property is determined eligible for listing on the National Register, then the Advisory Council on Historic Preservation must be afforded the opportunity to comment on any Federal project that may affect it. According to ODOT, the "time saving will be incalculable" from having information on potentially eligible sites readily available.

Stewardship and Streamlining

Transportation decisionmakers frequently cite Section 106 as one of the most cumbersome and contentious issues in the environmental review process. Yet according to ODOT officials "it's rare for Section 106 to be a hot button issue now—in other words to determine whether or not a project will be delivered on time." This finding is particularly striking when one considers the vast amount of cultural resources in the State; Ohio ranks third in the country in the number of National Register listings with approximately 3,600 listings for buildings, sites, structures, objects and districts, including over 55,000 contributing properties. ODOT credits good working relationships and the GIS as factors that have contributed to its relatively smooth track record in complying with Section 106 requirements.

By reducing conflict in what can frequently be a difficult area of the environmental review process, the GIS results in time- and cost-savings for ODOT. With the GIS, detailed information can be queried or "boiled down" and displayed visually for use in the early stages of transportation project planning. This process enables OES staff to electronically plot archaeological and historic site locations and make early evaluations of potential impacts in a transportation project area.

The GIS/MAPIT program also eliminates the time ODOT staff used to spend driving to the Ohio Historical Preservation Office to manually look through files to extract and copy information. ODOT has already saved hundreds of hours in data collection, thus leaving time for staff to spend on other activities and enable project reviews to move more quickly. Furthermore, consultants under contract with ODOT must use the same data search process and the GIS allows them to access the data at OSHPO via terminals at their own office, or even from ODOT. Lastly, the GIS affords ODOT staff the opportunity to screen an area for obviously significant cultural resources and therefore devote resources and detailed analysis only to the sites that really warrant it.

In addition, the very process of creating the shared GIS database resources has strengthened interagency working relationships and contributed to a high level of trust between ODOT and OSHPO. By meeting regularly with the OSHPO, ODOT has benefited with more timely review and approvals and a better understanding of each agency's goals and objectives in regards to the environment. The commitment to maintain the knowledge level of ODOT's OES and OSHPO staff is a critical element of the GIS/MAPIT process and is an example of environmental stewardship and streamlining between agencies. In addition, OSHPO uses these data jointly with ODOT to develop additional models and programmatic approaches to address cultural resource issues on a statewide basis. This program illustrates trust- and relationship-building efforts between ODOT and the OSHPO.

Lessons Learned

The very process of developing a shared GIS database and tools for analysis, planning, and visualization can be a streamlining activity. While ODOT and OSHPO have historically had good working relationships, ODOT's decision to finance OSHPO's development of a historic and cultural GIS database has made that relationship even stronger. Other States may consider the development of a GIS as a way to build trust and strong working relationships with partnering agencies, which contribute to a more efficient NEPA process.

Adapt your system to suit users' needs. A major initial stumbling block for ODOT was whether to use the GIS software that SHPO staff was accustomed to using or whether the system should be based upon ODOT's software. By developing a web portal interface between the two, the staff at both agencies was able to continue to do their work in the GIS environment in which they are most comfortable.

Consider an alternative institutional home for your GIS database. Sharing ownership of the GIS between OSHPO and ODOT created a win-win situation for both agencies. ODOT is now able to access the information they need more quickly and in a form that is easier to analyze, manipulate, and display. With the financial contributions from ODOT, OSHPO is now better able to catalogue and advocate for the protection of cultural resources in the State.

GIS for Environmental Stewardship and Streamlining
An Overview of State DOT Practices

APPENDIX E: WASHINGTON CASE STUDY

PRACTICE TITLE: *The Environmental GIS Workbench*
CONTACT: *Elizabeth Lanzer*
EMAIL: *lanzere@wsdot.wa.gov*
PHONE: *(360) 705-7476*

Introduction and Background

An obstacle many State Departments of Transportation (DOT) face when conducting environmental assessments is collecting the necessary environmental information from various resource agencies. Washington State DOT's (WSDOT) Environmental GIS Workbench (Workbench) demonstrates the power of a GIS database in managing disparate sets of information and streamlining inter-agency cooperation.

The Workbench is a custom application built to assist WSDOT staff in accessing over 60 layers of environmental, natural resource management, and transportation data. The WSDOT Environmental Information Program works with Federal, State, and local agencies to maintain a collection of the best available data for statewide environmental analysis.

The Workbench provides WSDOT staff with a tool to locate proposed transportation projects and display relevant environmental data themes for that location or route. Prior to the Workbench, users seeking this data had to navigate through a difficult environment that required them to know detailed information about scale and agency management of the data. The Workbench provides a more intuitive method to locate this information.

> "The intention [of the Workbench] is to reduce the amount of training and the learning curve that is presently needed by new GIS users to access to existing data, thereby improving the efficiency and the quality of the review process."
>
> –*Environmental GIS Workbench 2.0 Users Guide*[20]

The Workbench serves as both an internal review system and as a tool for multi-agency coordination. The Workbench has significantly improved the way transportation decisions are made in Washington State. Yet, notably, the application only took one year to become operational.

The Challenge

The Workbench grew out of an identified need to coordinate with local agencies on priorities for transportation system activities. The primary objective for the Workbench was to make project scoping (particularly budgeting and scheduling) more efficient by reducing the need to gather data from many agencies for a given project. Prior to the Workbench, gathering the data needed to begin the scoping process was a time-consuming, laborious process requiring consultation with many State and local agencies. Data redundancy among various

---

[20] Available at http://www.wsdot.wa.gov/environment/envinfo/docs/userdoc_EWBv20.pdf

GIS for Environmental Stewardship and Streamlining
An Overview of State DOT Practices

State agencies and the need to improve data quality were additional reasons that precipitated the development of the Workbench.

The Details

The Workbench is an ArcView 3x extension that was written in Avenue code. To users, the Workbench appears as an extra button (the "blue button," as WSDOT staff refer to it) on the top tool bar in ArcView. The Workbench is a user-friendly navigation system to locate GIS data on four servers:

(1) A GIS data server, which retrieves data layers via the WSDOT corporate server;

(2) A DOT Applications Server, which uses a Dynamic Link Library to convert Milepost values to their Accumulated Route Mile value;

(3) The State route (SR) View Server, which retrieves SR View images (digital images of the State Highway System from video log files); and

(4) A Spatial Database Engine (SDE) server (SQL Serve based) which provides geodatabase access.

The Workbench provides a basemap of Washington State and the project site. From there, users can add a proposed transportation project to the map along with relevant environmental data of the area, such as parks, wetlands, superfund sites, wells, and Natural Heritage sites. The user also has the capability to add additional layers, including SR View Images and digital orthophotography and embed these images into maps. Once the desired layers have been added to the map, users can create buffers of the transportation project—for example, at 20, 40 and 60 feet—to observe what impacts an alternative may have on local environmental features. With the Workbench, users are also able to easily view metadata and identify potentially environmentally sensitive areas. According to WSDOT, "users now have the capability to build custom maps in real time, perform spatial analysis, and create hardcopy prints of their work."

Data: Data for the Workbench comes from Federal, State, county, local, and tribal agencies; and, academic institutions. For example, the Washington Department of Fish and Wildlife has agreed to provide the Environmental Information Program with data on sensitive biological species and also with periodic updates of those data.

The Environmental Information Program "takes all the data [they] can get," and therefore, the data that they receive ranges considerably in quality. Upon receiving new data, the GIS staff standardizes it, "cleans up" the metadata, and populates it to the Workbench. The GIS staff then sends out an email to all Workbench users describing what has been added and the types of uses for which the data may be appropriate. The GIS staff trains general users to look carefully at the available metadata and the scale of the data to determine if the data are appropriate for their purposes.

Despite a State legacy of charging for spatial data, most agencies in Washington are now willing to share data. However, WSDOT staff notes that some local agencies are still struggling to fund their GIS effort by charging GIS data. Occasionally, WSDOT has traded imagery from their aerial photography center for data from other agencies.

GIS for Environmental Stewardship and Streamlining
An Overview of State DOT Practices

Training: A six-hour training session for WSDOT staff is advertised on the Workbench website.[21] The course is intended to train WSDOT staff in the fundamentals of working with ArcView, the Workbench tool, and the environmental data available through the tool. While the GIS staff focused initial training for the Workbench on project scoping, additional training is being developed for the expanded uses of the application.

The website also provides users with contact information for the "Environmental GIS Workbench Support Person," and a method for WSDOT staff to request training in the Workbench. A users guide geared for WSDOT staff is available on the site.

Stewardship and Streamlining

WSDOT characterizes the Workbench as an excellent return on their investment. The Environmental Information Program finds that training staff to use the Workbench to make basic maps and retrieve spatial information saves GIS staff time in the long-run as other staff no longer rely on them for these tasks. As a result, the Environmental Information Program staff can focus on managing and collecting new data. Another reason why the Workbench has been cost effective is because it is user-friendly and, therefore, has not required staff to undergo extensive training to acquire many additional skills.

The impact that the Workbench has had on the scoping process is significant. The GIS has been of primary importance in saving research time. Staff are no longer forced to "hunt down" data because the information is readily accessible. While the Environmental Information Program staff has not conducted a formal evaluation, they do note anecdotal evidence of time and cost savings. For instance, one project manager reported that the scoping process, which used to take him eight hours, could be completed in two hours with the Workbench. This anecdotal evidence suggesting that the Workbench reduces research time by 75 percent indicates that the Workbench has already paid for itself.
As the Workbench adapts to new uses within WSDOT, new advantages of the system are becoming evident. For example, the permitting staff has become more efficient in the field because they know what to expect based on the quick maps they are able to produce in their offices.

The Workbench has also been useful on specific types of issues. WSDOT credits having a standard dataset available and ready to use as a particular asset in negotiating disputes involving endangered species, a strongly contested issue in Washington State. In particular, WSDOT is using the Workbench to set acceptable limits and times on construction scheduling due to concerns about endangered species.

In addition, central management of data by expert staff improves data quality and reduces data redundancy throughout the State.

Lessons Learned

WSDOT's experience with the Workbench suggests several lessons that may aid other States considering the development of a similar GIS:

---

[21] Available at http://www.wsdot.wa.gov/environment/envinfo/egwbhome.htm

GIS for Environmental Stewardship and Streamlining
An Overview of State DOT Practices

Training investments yield a high return. Investing the time up-front to train staff in simple mapping techniques saves staff time by reducing the need to "hunt down" necessary data. In the case of the Workbench, the application also conserved the time of the GIS staff who, freed of obligations to create simple maps for staff, can now devote more time to assuring data quality and seeking out new sources of data.

Consider evaluation measures from the outset. WSDOT notes the importance of collecting baseline information about the tasks that a given GIS application seeks to improve. In the case of the Workbench, WSDOT wishes that they had data on how long it took staff to complete scoping work before the application was developed. This baseline information would make possible a more robust evaluation of the Workbench.

Build in flexibility to your GIS. WSDOT recommends that if States are thinking of developing an application for a narrow use—such as project scoping—that they get a sense of how other users might also use a GIS. Designing the GIS with as thorough a sense as possible of future uses would obviate the need to retrofit the application later.

Consider the scope of your GIS. A unique feature of the Workbench is the way the Environmental Information Program carefully limited its scope. For instance, the department has no plans to make the Workbench a publicly accessible GIS tool. One reason is legal: WSDOT is not the originator of the vast majority of the data and so cannot release this information to the public. However, the Environmental Information Program has also made the decision to focus on tools that can be useful to WSDOT staff in project delivery. As a result, they have not chosen to devote staff time and resources to the development of an Internet Mapping Server (IMS). They instead participate in a statewide GIS consortium, which is planning on developing an ArcServer web service.

Next Steps

A new tool is being developed for the Workbench that will incorporate land use land cover, geology, soils, wetlands, floodplains, steep slopes, critical aquifers, parks, hydrography and existing transportation infrastructure as inputs to a spatial model that will generate a mitigation risk index. The mitigation risk index estimates the cost effectiveness of mitigation highway impacts within the ROW. The more negative the resulting value, the stronger the implication that existing conditions and characteristics of the proposed project area will have a difficult time creating on-site, in-kind mitigation. Positive values indicate that mitigation is feasible within the ROW. A "perfect" score of 1.0 indicates conditions favorable to minimizing mitigation costs. The GIS tool will have an interface for users to input project locations, answer a few questions about project activities, and review the list of data that will be used. Once the inputs are validated, the model runs and provides the user with some statistics and a related explanation regarding mitigation issues.

While the initial purpose of the Workbench was to support project scoping, planning, and engineering, permitting staff have also become interested in using the application. Because of this growing interest, the scope of the Workbench itself is expanding. The next generation Workbench is expanding the utility of the application beyond solely environmental purposes to a new focus on maintenance and transportation planning. Elizabeth Lanzer, the Environmental Information Program Manager, describes this as an effort to expand the usefulness of the application "to meet all the GIS business needs of WSDOT."

GIS for Environmental Stewardship and Streamlining
An Overview of State DOT Practices

APPENDIX F: MINNESOTA CASE STUDY

PRACTICE TITLE: *Mn/Model*
CONTACT: *Elizabeth Hobbs*
EMAIL: *elizabeth.hobbs@state.dot.mn.us*
PHONE: *(651) 296-9243*

Introduction and Background

Since 1996, the Minnesota Department of Transportation (Mn/DOT) has been developing an archaeological predictive model, Mn/Model, to avoid impacts to archaeological sites throughout Minnesota. An archaeological predictive model is a tool that indicates the probability of encountering an archaeological site anywhere within a landscape. The probabilities of finding cultural resources sites are reflected in sensitivity maps. These maps, which usually contain three zones: a high sensitivity area where archaeological sites are most likely, a medium sensitivity area where sites are less likely, and a low sensitivity area where sites are unlikely, are beneficial for transportation and land-use planning. If construction projects can be modified to avoid areas where archaeological sites are predicted to occur, the result is better solutions.

Mn/DOT developed Mn/Model to aid in the avoidance of impacts to archaeological sites throughout Minnesota. Mn/Model used a combination of GIS-based tools and statistical modeling procedures to map the potential for pre-1837 surface archaeological sites in Minnesota; after 1837, white settlement began and settlement patterns were altered by a new set of cultural and economic factors. For the post-1837 period, historical maps and recorded history provide more accurate information than models.

Mn/DOT uses the predictive models for site avoidance and survey design. The results of Mn/Model are incorporated into the earliest phases of project planning, making transportation planners aware of the possible locations of pre-contact archaeological sites. Mn/Model, which is easily improvable as new archaeological and environmental data become available, allows planners to prepare alternative avoidance design scenarios, when possible, and to budget for survey and mitigation costs and time when avoidance is not possible. Mn/Model also helps prepare budget and schedule estimates allotted for individual projects and longer range management activities.

The Challenge

Mn/DOT must conduct archaeological surveys pursuant to Section 106 of the National Historic Preservation Act of 1966. Mn/DOT hires professional archaeologists to conduct these surveys. Before Mn/Model, determining where to survey depended on the experience of the professionals involved and changed with personnel. Often, in Mn/DOT's view, this approach resulted in more surveys than necessary. Mn/DOT, frustrated with the time and costs associated with survey and subsequent review, decided to develop a tool that would enable staff to discern where archaeological resources were likely to be found, and thus where surveys were needed most.

The Details

In 1995, with Federal Highway Administration (FHWA) funding in place, development began on the Mn/Model. The goal of the project was to use GIS and statistical analysis to produce

GIS for Environmental Stewardship and Streamlining
An Overview of State DOT Practices

archaeological predictive models that could be replicated by anyone using the same data and following the same procedures. The aim was that these models be accurate enough to predict 85 percent of known archaeological sites without designating more than 33 percent of the State's area as high and medium site probability.

When Mn/DOT first began developing GIS infrastructure at Mn/DOT, environmental data were not a priority. The first task for Mn/Model GIS staff was to develop a statewide database of environmental and cultural resource data in GIS format. These data served as the starting point from which Mn/DOT could begin focusing efforts on more ambitious analyses. The model, which was completed in 1998, is the most extensive high-resolution (1:24,000 scale or 30 meter cells) archaeological model ever created. Previously, similar models had focused on small areas such as parks. A location's sensitivity for archaeological site is based on a statistical analysis of more than 40 spatial environmental variables, such as terrain, proximity to water, slope, and vegetation. Models were developed independently for 24 ecological regions within the State. The model's results are displayed as a map, with each of the 30-meter cells classified as having a low, medium, or high potential containing an archeological site. Areas that have not been adequately surveyed, and therefore lacked sufficient archaeological data to model accurately, were classified as "unknown." The "unknown" classifications have helped Mn/DOT determine where surveys are needed because of lack of information, rather than potential for archaeological sites.

The model results have been used to suggest project alignments or modifications that reduce the potential for impacts on cultural resources and reduce the need for surveys. This information has allowed Mn/DOT to expedite project clearance, reduce costs, and do better job of protecting cultural resources.

Stewardship and Streamlining

With Mn/Model, fewer site surveys are necessary, saving Mn/DOT time and money. Surveys are now targeted to locations where there is a high potential for sites or where there is a lack of cultural resource information. Projects can also be reviewed more quickly. For example, currently during the NEPA process, three people are required to review historical properties, while only one is dedicated to reviewing archaeological sites.

Within two years Mn/Model repaid its investment with survey and mitigation costs savings alone. Total cost savings over the first four years of Mn/Model use have reached $3 million per year.

In addition to these savings, Mn/Model, as indicated on the Mn/Model website:[22]

- Allows Mn/DOT's Cultural Resources staff to clear more projects per year;

- Reduces the number of Memoranda of Agreement required;

- Improves project turnaround time– some projects have saved 1 or 2 construction seasons in survey time alone;

- Reduces schedule and budget uncertainty by reducing "surprises";

- Reduces cultural resource disturbance, a sensitive issue with Native American communities – Since Mn/Model was run, no new mitigations have been contracted;

---

[22] www.mnmodel.dot.state.mn.us

GIS for Environmental Stewardship and Streamlining
An Overview of State DOT Practices

- Supports coordination among governmental organizations – Mn/DOT is providing Mn/Model and training in its use and interpretation to Minnesota's SHPO, the Minnesota Office of the State Archaeologist (OSA), and the Tribal Historic Preservation Offices the US Forest Service, the Natural Resource Conservation Service, one county, and a regional planning agency; and,

- Leads to the development of new GIS data analysis and modeling applications and standards. For example, Mn/DOT can analyze view-sheds to determine the visual impacts of highway or bridge projects on cultural and historic properties.

Lessons Learned

Perform extensive quality control on your GIS data. Digitize surveys and known archaeological site boundaries at the 1:24,000 scale. Acquire environmental data at the same scale. Check all data (including site locations, water bodies, vegetation data, etc.) for correct coding, missing values, and internal consistency. The models can be only as good as the data used to create them.

Update models when there is a critical mass of new data. As new and better environmental and cultural resource data become available, incorporate them into the modeling framework and run the models again. Since developing Mn/Model, Mn/DOT has acquired several new statewide environmental datasets, corrected locations of most archaeological sites, and added hundreds of newly discovered archaeological sites. When new models are run with these new data, planners can be more confident that their assessments of current projects are based on the best available data.

Field-test models. The results of new field surveys that are performed can be compared to a predictive model's results to test its dependability. This approach can give users confidence in the reliability of a model. For tests to be valid, it is important to conduct some surveys in the high, medium, low, and unknown site potential zones of the model. Surveying only the high potential zone would result in sites being found only in high potential areas, thus providing a false confidence in the model.

Consider Creation of a GIS Technical Committee. When first developing Mn/Model, Mn/DOT formed a GIS Technical Committee that included the Minnesota Department of Natural Resources, SHPO, and the Land Management Information Center. The Committee facilitated data sharing and provided useful guidance for developing the modeling database and models.

Next Steps

Mn/DOT is now in the process of collecting and developing better environmental and archaeological data. When this task is completed, the model will be run again and should produce significantly better results. Much of the improvement will come from the correction of location errors found in the archaeological sites database. Because known archaeological site locations are based on Universal Transfer Mercator (UTM) coordinates entered into a database, map interpretation or data entry errors sometimes put sites in the wrong section, township, county, or even outside the State. As time allows, staff are working to adjust the baseline information.

GIS for Environmental Stewardship and Streamlining
An Overview of State DOT Practices

There is now a movement towards integration among State agencies in Minnesota. Mn/DOT is working on an initiative to develop a common GIS database with SHPO and OSA. It has applied for funding to enhance the OSA website. Currently, the OSA site allows licensed users to search for information on the location of burial sites. With extension of the site, all archaeological sites within the State would be searchable and updateable. The online GIS application would allow counties access to archaeological data for their cities and licensed consultants a way to digitize updated field data. Such an application would take the data entry burden off of OSA, SHPO, and Mn/DOT.

GIS for Environmental Stewardship and Streamlining

An Overview of State DOT Practices

APPENDIX G: TEXAS CASE STUDY

PRACTICE TITLE: *GISST (GIS Screening Tool)*
CONTACT: *Sandra Allen*
EMAIL: *Sandra.Allen@fhwa.dot.gov*
PHONE: *(512) 536-5944*

Introduction and Background

While many Geographic Information System (GIS) tools are used to identify and display the spatial distribution of features on the landscape, the power of GIS technologies goes well beyond simple map-making. GIS can also be used to perform complex spatial analyses, such as prioritizing environmental resources along a transportation corridor or monitoring cumulative impacts to those resources. The partnership formed by the Texas Department of Transportation (TxDOT) and the Environmental Protection Agency (EPA) Region 6 to apply the GIS Screening Tool (GISST) to the National Environmental Policy Act (NEPA) process on the Interstate 69 (I-69) project exemplifies the potential of GIS to perform sophisticated analyses.

I-69—or the "NAFTA highway"— is a congressionally mandated, 1,600-mile interstate highway stretching the Mexican border in Brownsville and McAllen, Texas to the US-Canadian border in Detroit, Michigan. The goals of I-69 are to facilitate trade between Mexico, Canada, and the United States and to encourage economic development and transportation access to rural communities along the route. I-69 was also chosen in 2002 as a streamlining pilot project[23] under the Transportation Equity Act for the 21st Century, Section 1309.

Developed by EPA in 1996,[24] the GISST is a system that imposes a scoring structure on GIS coverages to inform decisionmaking and prioritize environmental protection. The system has many applications from evaluating soil permeability and erosion potential to assessing the cumulative impacts of swine feedlots in Oklahoma. However, FHWA and TxDOT are using the GISST as a screening tool for the NEPA Tier 1 Corridor Level decision. TxDOT uses the system to identify areas to avoid and to enable TxDOT decisions about where to concentrate resources for further studies at NEPA Tier 2. The GISST has been designed to better understand the potential significance of single and cumulative impacts and to facilitate communication of technical and regulatory data with industry, the public, and other stakeholders.

GISST improves the quality of the environmental review process, while also expediting its delivery in several ways. By explicitly establishing a clear rating system for environmental resources, the GISST makes the NEPA process more objective, and facilitates improved

---

[23] For a complete list of the priority projects, go to http://www.fhwa.dot.gov/stewardshipeo/pplist.htm.

[24] The NEPA Compliance/GISST Developers Team consists of Gerald Carney, Ph.D., Jeff Danialson (ACS), Dominique Lueckenhoff, Sharon Osowski, Ph.D., David Parrish, and Joe Swick.

GIS for Environmental Stewardship and Streamlining

An Overview of State DOT Practices

agency communication and environmental stewardship and streamlining. GISST has been hailed by Anne Miller, Director of EPA's Office of Federal Activities, as a national model for the "integration of GIS into an overall management systems process that seems to have transferability to other parts of the country."

The Challenge

The scale of I-69 means that there is a potential for considerable environmental impacts due to road alignment and construction; yet, in light of the degree of national attention given to this project, avoidance and minimization of adverse environmental impacts is paramount. TxDOT has developed a list of strategies to streamline the environmental review process and minimize environmental impact, including utilizing a tiered approach to NEPA, early cooperation and collaboration with resource agencies, internal training on streamlining strategies, and the use of mitigation banks. Another of TxDOT's strategies is to reduce field surveys by using GIS technology to identify priority resources. FHWA and TxDOT identified EPA's GISST as the best tool to promote this goal.

Upon selecting a tiered NEPA process for I-69, FHWA and TxDOT divided the Texas portion of the Interstate into 13 separate Segments of Independent Utility (SIUs). In Tier 1, TxDOT will complete a general Environmental Assessment for the entire project— a broad area that is approximately 20-50 miles wide and 1,000 miles long. After a preferred corridor has been identified, which may be between 2,000 feet to several miles wide, the second phase of the environmental study will include a detailed evaluation of potential impacts to the natural and human environments along each of the 13 SIUs. Tier 2 of the environmental study will identify a final route alignment within the preferred corridor that has the least practicable environmental impact. TxDOT will use the GISST and related QUANTM software package— developed by TxDOT consultants—in Tier 1 of the NEPA process to initially screen for critical environmental resources and to narrow the study area for Tier 2 analysis.

Another unique element of the I-69 alignment selection process is the partnership between TxDOT, the Texas Parks and Wildlife Department, and NatureServe,[25] a non-profit formed by the Nature Conservancy and others to collect and manage data about the distribution of critical species and ecosystems. Texas State law protects private property by prohibiting State agencies from releasing any information related to resources located on private land without the owner's consent. Because of these legal requirements for confidentiality, the Parks and Wildlife Department are the gatekeepers of resource data. They run QUANTM along the digitized I-69 corridor and buffer it to protect private property rights. While TxDOT is concerned about paying for overly buffered data from the Parks and Wildlife Department, data available from NatureServe can verify these data in lieu of full public disclosure.

The Details

The GISST evaluates environmental vulnerability and impact through the use of over 100 types of environmental resource and stressor criteria developed by EPA. The scoring

---

[25] For more information see http://www.natureserve.org/.

GIS for Environmental Stewardship and Streamlining
An Overview of State DOT Practices

structure consists of criteria based on available data sets and expert input.[26] These individual criterion scores can be computed and assessed for the base unit of interest (e.g., watersheds, facilities, or NEPA alternatives). The scoring structure is essentially a prioritization tool for environmental resources located in a given basemap unit. Users can prioritize criteria on a one to five scale, where one represents no concern and five represents a high level of concern. The GISST scoring criteria evaluate the potential for ecological, socioeconomic, toxicity, landscape, air quality, and water quality risk, as well as opportunities for pollution prevention. The above criteria can be applied to the basemap unit as a function of **area** (ratio of the project area to the geographic unit), **vulnerability** (characteristics of environmental features), and **impact** (specific activities which will occur on the site).

The utility of GISST is its mapping and analytical capabilities. GISST combines the collective technical assessments into a mathematical algorithm and uses "natural weighting" to identify and map environmental concerns. This evaluation can ensure that decisions are made by the environmental information or criteria characterizing that geographic area and not by an arbitrary assignment. The result of this analysis is a path of least impact for possible NEPA alternatives and identification of mitigation opportunities. Individual criterion and the sum of several criteria can be used to determine alignment alternatives. The summation of criteria can be used as a measure of potential cumulative effects. Traditionally, criteria and the final GISST scores were calculated on a watershed sub-unit basis, however because it is more appropriate to use the SIU as the base unit for the I-69 project, the GISST has been modified to allow users to analyze the data on a one kilometer grid. The QUANTM expands on the system by allowing for the analysis of irregular polygon features.

Cost - An Interagency Agreement was developed, signed, and funded (more than $100,000) to support EPA assistance to FHWA and TXDOT. In return, EPA has provided TxDOT with data covering a 1000-mile corridor that ranges from 20-50 miles in width.

Training - TxDOT tailors a QUANTM course by applying it specifically to the I-69 project. This one-week course offers certification for up to six engineers and trains them on how to update data and change constraints.

Stewardship and Streamlining

With GISST TxDOT hopes to significantly cut the NEPA Tier 1 processing time. This is largely because of the way GISST helps to identify large-scale critical areas through its screening capabilities. The relatively quick and easy screening process offered through GISST points out 'red flags' to prioritize the areas to avoid and where additional information and analysis is needed at NEPA Tier 2.

FHWA and TxDOT officials also credit the GISST with an increase in trust and cooperation between historically disparate agencies, such as the EPA, the US Fish and Wildlife Service, the US Army Corps of Engineers, the Texas Parks and Wildlife Department, and the Texas Commission on Environmental Quality. Gaining consensus on the data and the criteria by which to rank features results in less conflict and more credibility for the transportation

---

[26] See Chapter Three of the GIS Screening Tool (GISST) User's Manual, available at http://www.epa.gov/Arkansas/6en/xp/enxp2a3a.htm

GIS for Environmental Stewardship and Streamlining
An Overview of State DOT Practices

planning process. GISST also fosters better interagency cooperation by creating more consistency in the NEPA process by virtue of applying the same process to various decisionmaking points.

GISST can also help preserve institutional knowledge. As staff retires or moves to different jobs, knowledge of programs and regulations is lost. GISST criteria and scoring system capture this knowledge.

Of key importance for FHWA and TxDOT is that EPA endorsed the use of the GISST on the I-69 project. An important consideration that DOT officials must give to the development of GIS screening tools or models is whether or not EPA and other resource agencies will concur that the data is sufficient for the decision at hand Consulting with the resource agencies prior to and during GIS development assures the DOT that its analyses will be accepted.

While GISST presents several clear and tangible benefits, it is also important to discuss the potential drawbacks of the system. The most important of these is that the GISST is a screening-level tool only. It does not replace traditional risk assessment or field investigations. GISST can only point the user in the direction of where problems are likely to happen or where resources should be directed for additional studies. Other drawbacks with GISST concern its reliance on available data, equally weighting data with different levels of quality assurance, and combining databases with different coverage accuracy and precision (e.g., county-level versus census block information).

Next Steps and Lessons Learned

Over the next year, FHWA and TxDOT plan to apply GISST products to the new Trans-Texas Corridor Plan. An additional $50,000 will be provided to the EPA Region 6 Office using I-69 streamlining funds to fund new applications of the GISST.

Lessons from TxDOT's use of GISST include:

Do not reinvent the wheel. If a resource or cooperating agency has already developed a GISST that can be easily tailored to the NEPA process for highways, then State DOTs can save a lot of time by forming partnerships with them. In addition to reducing the duplication of efforts, these partnerships can also reduce interagency conflict, which can further improve and expedite the environmental review process.

GIS can be adapted to be an effective screening tool for environmental alternatives. By identifying critical resources along a transportation corridor that warrant focused investigation, GIS technologies can save DOTs time and money by focusing detailed analyses only on the issues and areas that truly warrant them.

GIS technologies can assist in assessing cumulative impacts. Assessing cumulative impacts is an area of the NEPA process with which most agencies struggle. Utilizing GIS to catalogue and assess cumulative impacts over time to a given resource can greatly inform the decisionmaking process and reduce workload and stress to agencies that struggle with this complex issue.

GIS for Environmental Stewardship and Streamlining
An Overview of State DOT Practices

APPENDIX H: VIRGINIA CASE STUDY

PRACTICE TITLE: *Enterprise GIS, Natural Heritage Resource Database, and CEDAR*
CONTACT: *Dan Widner (Natural Heritage Resource Database) and Angel Deem (CEDAR)*
EMAIL: *Dan.Widner@VirginiaDOT.org; Angel.Deem@VDOT.Virginia.gov*
PHONE: *(804) 786-6762*

Introduction and Background

GIS and information technology are rapidly changing the way the Virginia Department of Transportation (VDOT) conducts business. After several years of developing in-house GIS capabilities, VDOT now boasts an Information Technology Application Division employing 120 people (State employees and consultants), and the agency is leading GIS efforts in Virginia to catalogue transportation and natural resource data for use in transportation geospatial applications. VDOT officials expect that GIS will provide more than data management and map-making capabilities; they believe that GIS can change the business process within the DOT, fostering better communication and ultimately better decisionmaking.

VDOT's GIS4EST work consists of several discrete projects. For instance, VDOT's Information Technology Application Division has assembled transportation and environmental data from internal DOT Divisions and resource agencies into one data repository: VDOT's Enterprise GIS. The Environmental Group previously stored transportation and natural resource data among 60-70 scattered databases and spreadsheets. These redundant systems represented a sizable waste of staff time and effort. With the Enterprise GIS, environmental staff can access spatial data at their desktops instead of searching through paper files or myriad, unintegrated systems.

VDOT also financially supported the creation of a Natural Heritage Resource Database, which was developed by the Virginia Natural Heritage Program (VNHP). The development of the natural heritage resource database has ensured that VDOT staff has easy access to reliable data essential to the NEPA process.

However, the GIS4EST application that represents a re-engineering of business processes is Comprehensive Environmental Data and Reporting (CEDAR), a spatially enabled project management tool. CEDAR is a workflow application with a spatial component that provides project management capabilities, a mechanism to track project progress, and a way to improve internal, interagency, and consultant communication. The project management capabilities of CEDAR enable users to notify users in other groups or agencies with questions and concerns, track projects, send email notification, and assign roles and responsibilities.

Challenges

Prior to the development of the Enterprise GIS, VDOT was essentially paying consultants to collect similar data each time the DOT undertook a project. According to one DOT official, consultants were even lamenting the fact that VDOT did not collect these data for future use. Furthermore, as noted above, internal systems to track data were spiraling out of control.

GIS for Environmental Stewardship and Streamlining
An Overview of State DOT Practices

DOT officials identified the development of the Enterprise GIS as the best solution to these data management problems. VDOT has now developed standards to ensure that consultants provide data in a format that can be easily integrated into the Enterprise GIS.[27]

The Details

The Enterprise GIS - Six years ago, VDOT began the development of the Enterprise GIS—a data repository that is 95% based on ESRI products, such as ArcView, ArcINFO, Oracle databases, and ArcIMS. VDOT asks that these agencies supply updated data every three to six months to the DOT.

Initially, VDOT officials expected that they would need to offer extensive training to build staff GIS competencies. However, the Information Technology Application Division has developed a web-based tutorial, provided a service to answer questions via email and phone, built a robust help into the application, and made the metadata FGDC compliant. The Division has thus been able to screen out the most frequently asked questions, and the DOT staff is able to teach themselves the fundamental skills necessary for basic tasks. The selection of a web based GIS approach has also meant that VDOT has not needed to purchase as many copies of expensive GIS software.

The Information Technology Application Division does offer training in special case situations for "power users," such as the Environmental Division staff. They also have organized a "road show" demonstration tour, in which they traveled from Richmond to the District Offices.

CEDAR, *Comprehensive Environmental Data and Reporting* - The first phase of CEDAR will provide a tool for project documentation and management for in-house users, as well as a way to track and monitor workflow. The first stage culminated in a statewide training in the summer of 2004. Once security issues are resolved in the second phase, the Information Technology Application Division will implement web accessibility so that resource agencies and environmental consultants can also use the system. VDOT expects that providing access to resource agencies and consultants will greatly enhance communication in the NEPA process.

VDOT has developed and enabled CEDAR for use on all types environmental projects, including those that receive Federal funding and are required to be submitted to NEPA, as well as those that are fully funded by the State. While the latter projects are outside of the NEPA process, they are still required to undergo a State environmental review process that requires agency consultation.

Natural Heritage Resource Database - VDOT has also formed a partnership with the Virginia Natural Heritage Program (located in the Department of Conservation and Recreation) to develop a spatial database of natural heritage resources. VNHP—whose mission it is to identify, protect, and preserve Virginia's biodiversity—did not have the financial resources to fully develop a GIS to catalogue and monitor these resources. However, because of a shared need for these data between the two agencies, VDOT entered in an agreement with VNHP to fund the comprehensive development of such a database that meets the needs of both VNHP and VDOT. Once the agencies agreed that the

---

[27] These standards are available at
http://gis.virginiadot.org/VDOT_Geo_Spatial_Data_Delivery_Recommendations.pdf.

GIS for Environmental Stewardship and Streamlining
An Overview of State DOT Practices

basic structure of the first iteration of the database would consist of comprehensive conservation sites coverage and thorough metadata, VDOT provided $119,000 for staff and $4,000 for computers. A Memorandum of Agreement outlined the terms of use of the resulting natural resource data and ensured that VDOT would have no-fee access to the natural heritage resource database for five years.

Stewardship and Streamlining

The Enterprise GIS - VDOT officials credit the Enterprise GIS with improving interagency relationships. Because of the GIS, VDOT officials regularly sit down with staff from other agencies, such as the Natural Heritage Program and the Department of Game and Inland Fisheries. Once trust was established between the DOT and resource agencies about how to use and interpret natural resource data, VDOT found that these agencies became more willing to share these data.

CEDAR - While CEDAR is still in the development phase, VDOT is already reaping the benefits of the system on the Interstate-81 road-widening project, which will run the entire length of the State of Virginia. VDOT staff estimate that the geospatial data in CEDAR has enabled them to shave approximately 1,000 hours off the contract resulting in an estimated savings of $100,000. These savings have been realized because CEDAR has obviated the need for each consultant working on the project to go through the data collection and assimilation process. VDOT expects to see repeated savings through the use of CEDAR.

Natural Heritage Resource Database – As expected, the agreement between VDOT and VNHP to develop a more comprehensive natural heritage resource database has proven mutually beneficial. The database has provided VDOT with easy access to data that was previously difficult to locate, enabled regional visualization of resource distribution, simplified decisionmaking, and created the ability to streamline project review procedures. For VNHP, advantages include a reduction in the volume of projects to review, an enhanced ability to respond to problem projects, and the database itself, which enables the agency to fulfill its mission more effectively.

Lessons Learned

Consider the appropriate sequencing for GIS development at your agency. According to a VDOT GIS official, there is a natural development process that State agencies undergo before arriving at a fully functioning and robust GIS. This process can be divided into two crude stages.

In the first phase, data collection is the main focus of attention. Initially, State DOTs or other State GIS coordinating bodies should bring people together around a common need or interest and focus on the development of communication strategies and building trust. To accomplish the latter, State DOTs could consider funding the development of spatial databases at resources agencies with scarce resources.

In general, once common data interests have been identified and communication between agencies has been established, the second phase is the time to push the actual GIS technology. At this point, DOTs should ensure that they have sufficient forward thinking and

GIS for Environmental Stewardship and Streamlining
An Overview of State DOT Practices

innovative GIS staff who thoroughly know the technology, including effective security measures and where the technology is heading in the future.

Develop your GIS via many modest pilot projects. It is important for States to divide GIS tasks into small pilot programs to gradually lead people down the path. State DOTs can use small projects to demonstrate the value of GIS to internal staff, resource agencies, consultants, and the public before they undertake more ambitious projects.

Identify GIS champions at high levels. Because of institutional barriers, one or two highly positioned champions can make a big difference in developing a robust system. Virginia was fortunate to have a Transportation Commissioner who used to work for an environmental consulting firm. Other States may not be as lucky; officials may need to seek out and cultivate these champions.

Consider ways to evaluate the effectiveness of your GIS. Developing evaluation measures is essential to justifying GIS project and ensuring long-term funding. To evaluate these projects, a VDOT official recommends documenting consultant fees as a fraction of the total project cost. While it might cost $20/hour to hire a GIS analyst, it may cost a great deal more to hire someone to build a database. Apply these factors to the number of hours of consultant time to come up with estimate of time and money saved. While future savings may be difficult to estimate once consultants stop including data collection and assimilation time into the scope of work, this may merely be a positive sign that true cost savings are being built into the process.

GIS for Environmental Stewardship and Streamlining
An Overview of State DOT Practices

APPENDIX I: FLORIDA CASE STUDY

PRACTICE TITLE: *The Environmental Screening Tool*
CONTACT: *Peter McGilvray*
EMAIL: *Peter.McGilvray@dot.state.fl.us*
PHONE: *(850) 414-5330*

Introduction and Background

Environmental regulations in Florida are more stringent than those implemented by the Federal government. The more prohibitive nature of Florida's environmental laws has helped increase support for technologies that can examine and evaluate environmental needs with high precision, accuracy, and speed. Recently, the Florida Department of Transportation (FDOT) recognized the need for a comprehensive interagency cooperation strategy in order to help facilitate the development and progress of environmentally sound transportation projects.

FDOT, along with the Federal Highway Administration (FHWA), joined in a cooperative effort with all of the Federal and State resource agencies with which FDOT works in order to redesign the planning, permitting, and project review process. The resulting Efficient Transportation Decision Making (ETDM)[28] process, which links transportation, land use, and environmental resource planning, has allowed for more efficient and effective incorporation of environmental data, project review, and technical assistance into transportation projects. Florida's ETDM has helped facilitate early and interactive involvement of all involved resource agencies, promoting the delivery of better and more consensus-based environmental outcomes.

GIS is integral to the ETDM process. FDOT has designed a GIS application that allows partnering agencies to electronically share data, compare analyses, and comment on proposed alternatives throughout the environmental review process. As a result, FDOT expects more efficient and effective environmental stewardship, along with considerable reductions in delays, project changes, and challenges associated with project development, permitting, and consultation. The process is expected to improve the quality of decisions and environmental investments.

The Challenge

The ETDM Process emerged out of an identified need to create a standardized and streamlined method for resource agencies to collaborate with transportation agencies on the environmental review process for transportation projects. The primary objective of the ETDM is to streamline the National Environmental Policy Act of 1969 (NEPA) review process by initiating early involvement and allowing for proposed alternatives to be continuously evaluated and updated.

---

[28] For additional information about ETDM, see http://etdmpub.fla-etat.org/.

GIS for Environmental Stewardship and Streamlining
An Overview of State DOT Practices

The Details

The GIS-based streamlining component of the ETDM is called the Environmental Screening Tool (EST), which is maintained by the University of Florida's GeoPlan Center[29] and its Florida Geographic Data Library (FGDL).[30] EST features an Internet-based application that is linked to an electronic database system. Users can use EST interface to view and comment on the results of GIS analyses related to the environmental impacts and requirements of proposed project plans and alternatives. In long-range planning, agencies can evaluate cumulative impacts on a project and system-wide basis. The agencies then are able to consider the interrelationship between land use, ecosystem management, and mobility plans with an integrated approach.

While previously, access to the EST was only available to members of an Environmental Technical Advisory Team (ETAT), groups formed specifically to complete the environmental review process, public access to EST went live in the Fall of 2004.[31] The public can now review projects at the same time as agencies and submit their comments and concerns to their Community Liaison Coordinator.

Each of Florida's seven districts has an ETAT, which consist of FDOT district staff and planning and regulatory staff from State resource agencies. During planning, ETATs can use the EST in an advisory manner, providing input on regulatory and planning priorities. ETATs can also comment on avoidance, minimization options, and mitigation options, allowing for a more accurate estimation of project costs.

During project development, the role of the ETAT changes from advisor to coordinator. ETATs use EST as a way to provide their input and technical assistance related to permitting decisions. EST is broken into five modules.

1. Project Input Module: The Project Input Module is the place where new transportation projects are placed into the database system (currently, over 200 projects prepared for analysis are in the system). EST provides three ways for new projects to be entered into the system. Existing GIS databases may be uploaded or transferred from the State Highway System (SHS) database. Data not already digitally available can be entered on-line using a digitizing utility. When projects are added or modified, EST automatically analyzes the proposed projects using prescribed criteria developed by the ETAT; for example, calculating the acreage of wetlands within the impacted area, and counting the number of known historical and archaeological sites in proximity to the candidate project.

2. Project Management Module: The Project Manager tool allows ETDM Coordinators and their project management teams to review project entries for completeness. ETDM Coordinators also use the information in the Project Manager Module to, notify ETAT representatives that projects are ready for review and summary.

---

[29] See http://www.geoplan.ufl.edu/, for more information.

[30] See http://www.fgdl.org/ for more information.

[31] Available at http://etdmpub.fla-etat.org/

GIS for Environmental Stewardship and Streamlining
An Overview of State DOT Practices

3. <u>ETAT Review Module</u>: Once ETAT Teams are informed of a project for review, the project moves into the ETAT Review Module. At this stage, the EST is used as a mechanism for project analysis, review, and monitoring. In the ETAT Review Module, a GIS analysis of a proposed project is made available to all the involved resource agencies. Here, resource agencies administer projects and see that all requirements, need, and comments are taken into account. The ETAT Review Module allows all involved groups monitor their responsibilities and concerns, while helping to ensure that project progress continues. This module also allows for transportation planners to investigate "What if" scenarios outside of the ETDM process – Metropolitan Planning Organizations (MPOs) and resource agencies can review projects that are not included in the long-range plan.

4. <u>Sociocultural Effects Module</u>: The Sociocultural Effects Module enables users to enter community and human environment characteristics into the database and record the sociocultural effects of projects. Each geographic district in Florida has a Community Liaison Coordinator (CLC). The CLC is responsible for sifting through the community characteristics and comments in order to capture a summary of community sentiments. The CLC can also use the Project Input Module to incorporate this information into the layers being prepared for analysis; community focal points, or areas with notable community opinion or concern, can be digitized into existing data layers.

   Resource agencies have 45 days (with a possible 15 day extension) to review a project and add comments in the ETAT Review and Sociocultural Effects Modules. After this time period, a project is re-evaluated by the ETDM Coordinator. Here, the ETDM Coordinator culls all of the GIS analyses and ETAT reviews in order to summarize the dialogue. The Coordinator has 60 days to create a Planning Summary Screen, and it is this report that is used as guidance to choosing a project alternative and to making certain that all environmental requirements have been addressed.

5. <u>Public Information Site</u>: The final EST module is the Public Access Site. Through the Site, the public can query the ETDM database to retrieve reports about project characteristics, agency comments, and GIS analysis results.

   Training - In order to support EST, training was required. EST training program consists of two delivery methods: 1) Hands-on training presented in a lab setting where the participants actively use EST's five modules and work through examples and 2) Web-based training classes. FDOT found the hands-on staff training most effective. The interactive module training helped create an interagency communication breakthrough. Trainees were provided applied experience, helping to demonstrate the user-friendly nature of EST. The outcome of FDOT's training was that other agencies began to view the GIS tool as useful and convenient. To date, FDOT has trained the ETATs in all seven districts. FDOT plans on expanding EST training by creating an easily accessible web-based curriculum.

Stewardship and Streamlining

Through ETDM and EST, FDOT anticipates a more effective and timely decisionmaking process that does not compromise environmental quality. Since the project development phase will be incorporated into NEPA with the EST, FDOT estimates that the time required to complete the environmental review process will be reduced on the order of several years.

GIS for Environmental Stewardship and Streamlining
An Overview of State DOT Practices

EST represents a shift in how all the stakeholders collectively communicate, interact, plan and manage transportation improvement projects. The collaborative nature of EST use may further enhance the expedited environmental review process FDOT anticipates. The Tool allows FDOT and collaborating agencies to visualize and address potential project flaws, while determining ways in which goals might be accomplished. EST promotes communication, concurrence, and early buy-in from all involved agencies, all crucial elements to speeding project implementation.

Another way in which EST is expected to streamline the environmental review process is by flagging and resolving concerns early in the process. ETATs will be able to focus on key environmental issues in their districts and will be better prepared to convey these issues to each other. The interagency communication and detailed reviews that EST supports should help ensure that ETAT concerns are noted and that any project disputes can be resolved before funding. In one case, a project was even removed from an MPO's long-range plan based on the concerns submitted by the ETAT, thus saving labor hours and project funds on an unworkable project.

Performance measures for achievement of overall time frames are being created though obtaining agreement on a proper measure of success has been difficult. FDOT has created a task force, which has a draft performance white paper currently under review.

Lessons Learned

EST helps planners identify major environmental and social issues early in the transportation planning process. Planners can then address these issues before additional time and resources are invested.

FDOT has faced many challenges during the development and institutionalization of the ETDM Process and EST. With the tools now in place, however, FDOT considers the efforts successful and looks forward to improving EST and GIS program in general. The following describes some of the lessons FDOT learned during the creation of EST.

Use Training to Demonstrate the Utility of GIS. Initially, FDOT found difficulty in communicating to Florida's 25 MPOs the potential benefits of EST. FDOT used training to overcome the reluctance of partnering agencies. Once these groups understood the simplicity of EST's three modules, "eyes were opened" to the beneficial impacts this Tool might have on the environmental review process.

The new challenge is encouraging collaborating agencies to move beyond simple buy-in of general EST use to giving the proper human resource and financial priority to the ETDM process. FDOT is exploring ways in which initial buy-in can be followed by continued advancement of operating agreements.

Look Beyond Initial Programmatic GIS Development Costs. The implementation of EST has not been an inexpensive endeavor. Before FHWA provided funding for the development of EST, FDOT had invested roughly $400,000. Since that time, approximately $1 million in State funds and $1 million in Federal funds have been directed towards the expansion of EST and FDOT GIS program.

GIS for Environmental Stewardship and Streamlining
An Overview of State DOT Practices

Despite this upfront cost, however, FDOT expects EST, and other GIS efforts, to begin to pay for itself through future money and labor-hour savings during NEPA review. FDOT also anticipates EST will help create savings beyond price by preserving the environment.

Develop Ways to Manage and Update Data. FDOT recognizes the challenges associated with updating GIS layers for EST. For this reason, FDOT developed innovative strategies to ensure accurate data is being used. For example, Florida is a rapidly urbanizing State, making it difficult to have current socio-cultural data. In order to overcome this challenge, FDOT is developing a Community Impact Assessment component for EST. For now, the Community Liaison Coordinator sifts through EST commentary and summarizes community sentiments.

Invest in Training to Yield High Returns. Investing the time up-front to train staff in EST not only sold them on them on its usefulness, but also holds the promise for long-term streamlining savings. While hands-on training may be most effective, explore web-based training guides to answer frequently asked questions and instruct on basic procedures.

www.ingramcontent.com/pod-product-compliance
Lightning Source LLC
Chambersburg PA
CBHW081901170526
45167CB00007B/3105